THE PRICE OF COAL

The Price of Coal

MICHAEL P. JACKSON

CROOM HELM LONDON

First published 1974

© 1974 by Michael P. Jackson

Croom Helm Ltd
2-10 St. John's Road London SW11

ISBN 0-85664-121-9

Set by Red Lion Setters, London
Printed and bound by
Redwood Burn Ltd, Trowbridge and Esher

CONTENTS

TABLES

ACKNOWLEDGEMENTS

I am grateful to members of staff and students in the Department of Sociology at the University of Stirling. While at Stirling I have learnt a great deal which has been valuable both directly and indirectly for the production of this book. Particular thanks are due to Mr. D. Wynn who read a draft of the book and made many useful comments.

Michael P. Jackson,
University of Stirling.

1. INTRODUCTION

The dramatic revival in the fortunes of the coal-mining industry in 1973 understandably excited considerable public attention.[1] Throughout the 1960s the coal-mining industry had, in a very real sense, been "fighting for survival". Coal consumption had fallen by almost 30 per cent between 1959 and 1969, while the number of men employed in the industry had declined by a staggering 55 to 60 per cent. The central question concerning the coal-mining industry in the 1960s had been how the contraction of the industry could be achieved with the minimum adverse social consequences.

In 1973, largely because of the restriction of oil supplies and the increase in oil prices, the climate changed. There was no longer a surplus but a shortage of coal; politicians pleaded with the miners not to take industrial action and interrupt supplies because of the catastrophic effect such action would have on the economy of the nation. The central question concerning coal-mining no longer dealt with the contraction of the industry; it asked how the supply of coal might be safeguarded and men attracted to work in the industry.[2]

It would be a mistake, however, to look at the events of 1973 in isolation for the coal-mining industry this century has suffered many other marked changes in fortune. Some have been just as dramatic as those of 1973. The downturn in demand in 1957, for example, took most observers by surprise. For the previous ten years coal consumption had increased regularly by about 3 million tons a year and in 1956 the National Coal Board had published its plans for the long-term expansion of output. Other changes may have been less dramatic but their consequences have been just as serious. For instance, although the reduction in demand in the 1920s followed four or five years of instability it nevertheless had a number of far-reaching results. The decline in demand contributed to the low profitability of the industry, the lack of willingness of the owners to invest, the attack on miners' wages, and the reduction in the number of men employed. One might also note the impact of the two world wars. Initially demand for coal dropped significantly and as a result many men left the industry to join the armed forces, but within a year or so the picture had changed; demand for coal increased and there was a shortage of miners. For the rest of both wars there was a constant fear that the shortage of coal would hamper the 'war effort'.

In general the coal-mining industry has not found it easy to adapt to such changes in 'market conditions'. Two factors have made matters particularly difficult. First, physical changes need to be planned a long while ahead. A coal mine can take anything from ten to fifteen years to reach full production from the initial decision to open the mine. Even the reconstruction of an old mine can take almost as long. It is not easy, therefore, to suddenly increase the production of coal, especially if a substantial or sustained increase is desired. It is not easy either to suddenly decrease the supply of coal, especially if there is a chance that one might want to increase the suply again at some later date. Once a mine has been closed down it might prove impossible to reopen it.

Second, mining has been, and to a large extent still is, a labour-intensive industry, with the result that major changes in the level of production inevitably necessitate changes in the level of employment. It is not, however, easy to recruit, nor, if one recognises certain social constraints, to dismiss considerable numbers of miners. The industry has always drawn the bulk of its recruits from pit towns or villages; isolated communities where there is little if any alternative employ-ment.[3] It may be possible to meet part of an increase in the demand for labour from traditional sources; either 'ex-miners' or miners' sons. There is, however, a limit to the extent to which this can be done. Certainly if the increase in demand is substantial then it is unlikely that it will be able to be met entirely in this way. In such an instance the industry will have to try to recruit men who have no connection with mining or the mining communities. Such a task is not an easy one, as was discovered during both world wars and in the late 1940s and early 1950s. In order to attract such men one not only has to persuade them to accept a 'different way of life' but one will also probably have to provide them with accommodation in the local area. When there is a decline in the demand for labour the problems are different, but no less difficult to deal with. If the decline is too sharp then you risk causing considerable social problems. If the men no longer required to stay in the pit village then they will probably be condemned to long periods of unemployment; if they move then they may find employ-ment, but lose the way of life to which they have been accustomed, be separated from their friends and relatives and so on. Further, if men do leave the mining community when unemployed then they are much less likely to return to the industry in response to an increase in the demand for labour at some future date. The crux of the problem is that with mining, to a large extent, occupational mobility necessitates geographical mobility.

One of the certal purposes of this book, therefore, will be to look at the changes that have taken place in the 'market position' of the industry, at the problems these changes have caused, and at the attempts made to overcome them.

2

Another related purpose will be to look at the reactions of various interested parties to these events. Have they merely accepted the inevitability of the changes or have they tried to fight against them; if they have tried to fight against them, how have they done so and what degree of success have they achieved? Further, what has determined the attitude of these different groups and to what extent has it been an immediate reaction or part of a long-term strategy?

It would be possible, of course, to look at the reactions of a wide range of groups and individuals, and to a certain extent this will be done. Attention, however, will be centred on three groups; the government, the employers (both private employers and the National Coal Board) and miners (plus their unions[4] – largely their main union, the National Union of Mineworkers and its predecessor the Miners' Federation of Great Britain). These groups have been chosen because it is believed that they have had a central role to play. Other groups, like consumers, have at times influenced events (in the middle 1930s, for example, major consumers played an important role in helping to find a solution to a conflict between the employers and the unions in the industry) but generally their influence has not been as crucial as that of the three groups mentioned.

The method adopted will entail looking at the relevant material, broadly in chronological order. Chapter 2 will be concerned with the period from the end of the nineteenth century to the beginning of the Second World War; Chapter 3 will look at the wartime period and the nationalisation of the industry; Chapter 4 will deal with the expansion of the industry in post-war years; Chapter 5 will centre on the decline of the markets after 1957; and Chapter 6 will examine the events of the early 1970s, including the two major industrial disputes of that period. In conclusion, Chapter 7 will draw together a number of the comments made in earlier chapters about the reactions of governments, employers and miners to the changes in fortune.

NOTES

1. The coal-mining industry has never lacked interested observers. During the 1960s, however, relatively little was written about the industry, certainly when compared to the inter-war years and the post-nationalisation period.
2. Recruitment was essential to maintain existing manpower levels; an increase in the level of manpower was not contemplated.
3. see, N. Dennis, F. Henriques and C. Slaughter, *Coal is Our Life,* Eyre and Spottiswoode, London, 1956, for a description of life in a mining community.
4. There have been, and still are, a range of different unions operating in the coal-mining industry. Almost all manual workers and many craftsmen, however, were before the Second World War represented by the Miners Federation of Great Britain, and have since the Second World War been represented by the National Union of Mineworkers. The other unions operating in the industry are largely managerial and white-collar unions (with quite small memberships) although certain craftsmen are represented by other unions.

2. TOWARDS A DEPRESSION

There is little doubt that for the British coal-mining industry the period immediately prior to the First World War was 'the age of success'.[1] Output between 1888 and 1913 increased dramatically, from 170 million tons to 287 million tons (see Table 1), while employment in the industry over the same period increased from 439,000 to 910,000 (about one-tenth of the employed male population). Profits benefited in a similar fashion; they rose from about £11 million in 1890 to £28 million in 1913 (see Table 2).

A review of the coal-mining industry published in 1913 concluded:

> There are few industries in this country in which prosperity at the moment is greater than it is in the coal-mining industry, or any in which the prospects over the year are brighter. Wages have been advancing at a more rapid pace than they have done at any period since the great boom of 1900-1901; employment was never more regular or plentiful; the profits earned by colliery companies are exceedingly good; and the foreign demand for export coal is so great that coal exporters and colliery owners are clamouring for greater dock accommodation and quicker dispatch.[2]

Coal, in fact, dominated the world's energy market. The western industrialised nations were almost totally dependent on it. Britain, in turn, was one of the world's leading producers of coal; she dominated the European if not the world scene. In 1913 she was producing half of Europe's and 20 per cent of the world's coal and she was the world's leading coal exporter. About 98 million tons or 34 per cent of the 287 million tons of coal produced in Britain in 1913 was exported.

There were, of course, even in 1913 some disquieting signs. Oil had started to challenge coal's dominance. More specifically the British industry was being challenged by the USA which had already overtaken her as the world's largest coal producer (the USA had on average produced 457.72 million tons per year between 1900 and 1913, although comparatively little of this had been exported). The British industry (fortunately like many of its competitors) was almost completely unmechanised and productivity was falling; in 1881, on average, each miner employed had produced 403 tons of coal a year but by 1911 the comparable figure had fallen to 309 tons. There were also disturbing signs on the labour front. Although Dron recorded that 'for a long period prior to 1912 the Mining Industry enjoyed a reasonable degree of tranquility in relations between employers and workers'[3] — tranquility that was, he suggested, the result of 'the sound common sense and moderation exhibited by both parties'[4] — all was not as well as might be imagined. In the early years of the twentieth century miners' wages fell sharply and although there was a recovery between 1905 and 1907 they had only barely climbed back to their 1900 level by 1913. In 1912 the tranquility referred to by Dron was shattered by the first major national strike in the industry. The strike was only settled by the passing of the Minimum Wage Act of that year (which determined that in future minimum wages were to be fixed, for each district, by specially appointed District Boards).

Nevertheless, despite these disquieting signs, there can be little doubt that to those in the industry, and to contemporary commentators, in 1913 the coal-mining industry's future had never looked brighter.

The War Years

The immediate effect of the war was to bring about a sharp decline in the demand for British coal. Britain, as a major exporting nation, suffered more than most from the loss of export markets during the early part of the war. Certain areas, like Northumberland, which were largely dependent on exports, suffered particularly badly. Production in Northumberland fell by about 12 per cent and employment by about 30 per cent between 1913 and the early part of 1915.

The immediate downturn in activity, though, was only a temporary phenomenon soon to be reversed. The decline in exports was more than compensated for by the increased demands of the navy and by the middle of 1915 Britain was facing a shortage rather than a surplus of coal. The shortage was to last throughout the rest of the war and to cause the government considerable concern. Coal was vital to the 'war effort'; not only was the home economy dependent on coal, but so was the nation's sea transport. The government was, therefore, increasingly forced to intervene in the coal-mining industry to protect the supply of this essential fuel, but because it was loath to involve itself in the industry's affairs its first steps were very tentative indeed.

The government's first action was to appoint the Coal Mining Organisation Committee. This was a committee without statutory powers, which was chaired by the Chief Inspector of Mines and composed of equal numbers of representatives of coal-owners and coal-miners. The Committee was charged to:

Inquire into the conditions prevailing in the coal-mining industry with a view to promoting such organisation of work and such cooperation between employers and workmen as having regard to the large number of miners who are enlisting for naval and military service will secure necessary production of coal during the War.[5]

These terms of reference give a hint about some of the industry's major problems. First, the industry was plagued by labour problems. The 'tranquillity' referred to by Dron, of earlier years, had disappeared. In 1914 nearly four million man days were lost as a result of labour disputes in the industry and in 1915, although the total fell, it was still well over one and a half million man days and as such accounted for over 56 per cent of all man days lost through stoppages of work in the UK in that year. Second, there was the problem of manpower. Large numbers of miners enlisted in the armed services in the early part of the war, partly as a result of the depression in the industry. Between July 1914 and February 1915, 134,186 miners left the industry, about 13 per cent of the industry's labour force. While this was acceptable for the first few months of the war when there was a surplus of coal, it was not afterwards when there was a shortage. Third, the industry faced significant organisational problems. Not only was action required in the mines themselves but it was also required to deal with the difficulties experienced over the distribution of coal. Distribution problems had been largely responsible for the shortage of coal in London in February 1915 and the consequent panic buying. It was hoped, then, that the Coal Mining Organisation Committee could find a solution to problems such as these and in so doing help to protect and if possible increase the supply of coal.

The Committee produced three reports. In their first report[6] the Committee recommended that four courses of action be considered. First, an appeal might be made to the miners to reduce absenteeism (at that time running at about 4.8 per cent). It was estimated that if successful this could result in a saving of 14 million tons of coal per year. Second, further investigation should be made to determine the situation with regard to exports. It was estimated that there had already been a reduction of 24 million tons in the amount of coal exported. Third, the industry was exhorted, wherever possible, to use more efficient means of production. Fourth, it was suggested that consultations should begin about the possible suspension of the eight-hour day in some districts. The length of the working day having been fixed at eight hours (excluding winding time) by the Coal Mines Regulation Act of 1908.

In fact, little action was taken on this report. No moves were made to abandon the 'eight-hour day'. At a conference held at the Home Office in September 1915 the miners' unions argued against any moves to suspend the eight-hour day and suggested that nothing should be done until the results of the efforts of owners and workmen to increase output in other ways could be evaluated. Similarly, no action was taken over absenteeism and the rate continued much as at the beginning of the war. There was, however, some improvement in the amount of time taken off by miners (through a reduction of holidays and the like) and some improvement in production by concentrating work on easily mined sections.

The Committee's second report[7] was published early in 1916. While they did not ignore the manpower issue in this report the committee clearly centred their attention on transport problems. Two areas were causing concern. First, there was concern that supplies of coal were frequently held up as a result of the shortage of rail transport. Second, there was the problem, particularly in South Wales, of the shortage of sea transport. This problem, the Committee suggested, had arisen largely because many British ships were no longer working in British waters. They had been attracted by the high rewards to sail between foreign ports. Action was suggested in both of these areas. To deal with the problem of rail transport the Committee suggested that a scheme for the pooling of rail wagons might be tried. To deal with the problem of sea transport the committee suggested first an inquiry into the extent and nature of foreign trade undertaken by British ships, followed, possibly, by restrictions on such trade.

Once again, however, little action was taken on the report. Although there was some improvement in the availability of sea transport no attempt was made to introduce a scheme for the pooling of rail wagons. Some action was taken, though, to deal with one of the problems highlighted in the Committee's earlier report. In June 1916 the government prohibited the recruitment of miners into the armed services and made arrangements for a number of ex-miners in the home services to be returned to the mines. It was estimated that under this latter arrangement the industry would benefit to the tune of 15,000 men.

In their third report,[8] issued in September 1916, the Committee recognised that some progress had been made. The government's measures to halt recruitment of miners into the armed services were clearly having a beneficial effect and the miners themselves, although they had failed to significantly reduce the level of absenteeism, had further increased the number of days worked; it was estimated that holidays had been reduced by 50 per cent from the pre-war level. They argued, however, that there was still room for improvement. They recommended that further attempts should be made to economise in the use of coal and that an extension might be sought in the local committees of workmen and managers that had been set up to deal with problems of coal production.

The Committee had spent much of their time dealing with the second and third of the problems mentioned in their terms of reference, manpower and organisation. They had a number of achievements to their credit in this area although they had been seriously hampered by their lack of statutory authority. Too many of their recommendations had simply been ignored. The Committee had done little, however, as regards the third problem, labour disputes.

The government, meanwhile, had taken some action itself. Following the report of a committee on the causes of the rise in the retail price of coal, the government had introduced the Price of Coal (Limitation) Act in July 1915. This Act attempted to limit the price of coal to four shillings a ton above the corresponding price on 30 June 1914. The miners had been particularly incensed by what they considered to be profiteering by the owners who had raised the price of coal steeply when demand had increased and so this helped to relieve some of the tension in the industry. It did not, however, remove all of the antagonism and it was to be this side of the problem that eventually forced the government to intervene in the industry's affairs in a more direct fashion.

The South Wales coal-field was the scene of a series of disputes in 1915.[9] In March 1915 the South Wales Miners' Federation put forward a demand for an increase in wages and special payments for miners involved in afternoon and night shifts. The owners, in reply, offered a 'ten per cent war bonus' but refused to consider the miners' specific claims. Consequently, the miners at the beginning of April gave three months' notice of their intention to terminate the existing agreement. It was not, however, until June that the government appeared to notice the gravity of the situation and realised that unless action was taken there would be a stoppage of work in the South Wales coal-field. The government sent two missions to talk to the miners' leaders. The second, containing the three Labour members of the government, persuaded the miners to accept a compromise settlement as 'a basis for negotiation'. The negotiations that took place, though, settled little, and by July the spectre of a strike in the South Wales coal-field loomed once more. On 12 July 1915 a delegate conference of South Wales miners passed a resolution declaring:

We do not accept anything less than our original proposals, and . . . we stop the collieries on Thursday next until these terms are conceded.[10]

9

The government's reaction was to prohibit a strike by proclaiming that the South Wales coal-field should come within the scope of the terms of the Munitions of War Act 1915 (which effectively made strike action illegal). The miners' reaction was instinctive and hostile. Incensed by the government's measure they came out on strike. It was quite clear that the government could do little to back its legal authority and after five days the government entered into negotiations with the South Wales miners' leaders and agreed that they would persuade the coal-owners to accept most of the miners' demands.

Discontent and disputes continued throughout the rest of 1915 and into 1916. First there was disagreement over the terms of settlement of the 1915 strike, then there was controversy over a government-appointed arbiter. Serious problems, involving the threat of strike action, did not arise again, however, until later in 1916. As before, the trouble began with a miners' wage claim.

At a special conference in October 1916 the South Wales Miners' Federation decided to press a claim for an increase in wages of 15 per cent. Their claim was based on the increase in the selling price of coal. They argued that the selling price of coal had risen by 1s. 11½d. a ton in the quarter ending June 1916 and another 7d. a ton in the quarter ending September 1916. Further, they argued that in order to test the owners' claim that the industry's costs had increased over the same period, the costs of production should be subject to a joint audit. The owners countered the miners' claim and argued that rather than be subject to an increase of 15 per cent wages should, in order to preserve profit margins, be reduced by 10 per cent. They also rejected the notion of a joint audit of costs of production, arguing that the present arrangements for verifying costs were satisfactory.

The government decided that they could not risk another stoppage of work in the area and in November 1916 announced that they were to take over the control of the mines in the South Wales area by an order made under the Defence of the Realm Act.[11] Early in December Asquith resigned as Prime Minister and Lloyd George, the new Prime Minister, let it be known that he would extend the control to the whole of the industry.

Under the terms of the Defence of the Realm Act the coal mines were brought under the general supervision of the Board of Trade. A new body was set up to supervise the industry, under the direction of the Controller of Mines, and assisted by an advisory committee of seven miners' and seven owners' representatives. The government through this machinery took control of prices, export licences, capital investment and the allocation of materials. It did not, however, take control of 'day-to-day' management, and the ownership of the industry was left in private hands. Essentially, then, it was a system of 'dual control' and was clearly a long way from nationalisation.

As a temporary expedient the government's actions had some success. Although the miners were far from happy about the 'half-way-house' situation – they argued for nationalisation – the government's intervention did improve labour relations somewhat. It also succeeded, to some extent, in improving the organisation and distribution of coal. On the other hand, coal production declined, from 256 million tons in 1916 to 248 million tons in 1917 and 227 million tons in 1918. Further, although the government undertook to guarantee profit levels this obviously failed to satisfy the mine-owners, and investment in the industry dropped dramatically, a fact which was to cause the industry many problems in the years to come.

After the War

The position of the coal-mining industry did not change radically with the ending of the war. Demand remained high and the government, despite opposition from the coal-owners, continued its supervision of the industry. The miners, for their part, determined that the time had come for the presentation of a new wage demand.

On 9 January the executive of the Miners' Federation decided to put forward a claim for a 30 per cent advance in wages. This claim was based on a comparison of miners' wages with two other factors. First, miners' wages had risen by 78 per cent since the beginning of the war (although this figure was contested; the Financial Adviser to the Coal Controller, Dickinson, produced figures which suggested that they had risen by 105 per cent) while the cost of living had risen by 120 per cent during the same period. Second, profits from the coal-mining industry had trebled since the beginning of the war, from £13 million to £39 million.[12] At a conference held at Southport in the middle of January, the executive claim was given full support, and was indeed increased by the delegates. It was decided that in addition to the claim for a 30 per cent increase in wages three further demands should be made:

(i)   special financial provision should be made for demobilised and unemployed miners;

(ii)   the 'Eight Hours Act' should be amended to become a 'Six Hours Act'. It was claimed that this would effectively mean on average a seven-hour working day because of the exclusion of winding time from the calculation;

(iii)   nationalisation of the industry. This was a reaffirmation of an earlier conference decision. It included a demand that nationalisation should be on the basis of a substantial degree of 'workers' control'.

After unsuccessful negotiations with the government over these demands, it was decided, at a special conference in the middle of February, to hold a strike ballot of the membership. The ballot showed an overwhelming majority in favour of strike action; 615,164 for strike action, 105,082 against. Strike notices were issued by the union due to expire on 15 March.

The government responded to this situation by proposing a Commission of Inquiry into the industry. The Commission, it was suggested, would have wide-ranging powers to examine all of the industry's problems, including the miners' claims. Further, a number of miners' representatives would be invited to serve on the Commission and it would be asked to present an interim report by 20 March. At a conference on 26 February the miners agreed to this formula and postponed strike notices until 22 March.

The Commission, when it was set up, had Mr. Justice Sankey as chairman, plus four representatives of the Miners' Federation, two members agreed by the miners and the government (Mr. R. H. Tawney and Mr. Sydney Webb), three government nominees and three representatives of the coal-owners.

The Commission acted quickly. It held its first meeting on 3 March and had concluded the examination of witnesses by 17 March. Its interim report[13] was published on 20 March. The speed with which the Commission worked, however, did not indicate any unanimity amongst the members. On the contrary, the Commission, in fact, produced three separate interim reports.

One report was presented by the four miners' representatives, supported by Mr. R. H. Tawney and Mr. S. Webb. Their report, as might be expected, broadly supported the miners' position. It recommended that the 30 per cent increase in wages be granted; that the 'Eight Hours Act' should be amended to become the 'Six Hours Act' and that because the individual ownership of collieries was 'officially declared to be wasteful and extravagant',[14] and

> in view of the impossibility of tolerating any unification of the mines in the hands of a capitalist trust . . . in the interests of the consumers as much as in that of the miners, nationalisation ought to be, in principle, at once determined on.[15]

The one issue over which the report did not give full support to the miners' case was the treatment of miners demobilised from the Army. It argued that rather than be treated separately it would be better if they were dealt with in the same way as men in other industries.

The second report was signed by the Chairman and the three government nominees. This report went part, but not the whole of the way, with the miners' case. It argued for an immediate amendment of the 'Eight Hours Act' to become the 'Seven Hours Act', allowing for the substitution of six for seven hours at a later date; wage increases of about 20 per cent; and changes in the method of ownership in the industry, although it did not declare itself specifically for or against nationalisation.

The third report was signed by the coal-owners' representatives. This report confined itself to the issues of wages and hours of work. It suggested wage rises of 15 per cent and the introduction of a 'Seven Hours Act'.

The government declared itself to be in favour of the second report, that produced by the Chairman and the government nominees. The miners were made an offer on this basis and after a ballot of their members it was accepted in the following month.

Having dealt with the immediate wages and hours issue the Commission reconvened to consider the structure of the industry in more detail. In the interval, however, significant changes in personnel had taken place. Sir Thomas Royden (one of the government's nominees) had resigned because of ill health and had been replaced by Sir Allan Smith (Chairman of the Engineering Employers Federation), and soon after the Commission began its second stage Mr. J. T. Forgie (one of the coal-owners' nominees) also resigned because of ill health and was replaced by Sir Adam Nimmo (another nominee of the coal-owners who had previously served as an adviser to the coal controller). The net effect of these changes was probably to increase the coal-owners' strength on the Commission qualitatively as well as quantitatively.

The Commission's final report was presented in June 1919.[16] Like the interim report, it was composed of a series of separate reports, this time four instead of three, although two of the reports, those by the Chairman and the miners' representatives, had a great deal in common.

Both the Chairman's report and that of the miners' representatives, plus Mr. R. H. Tawney and Mr. S. Webb, favoured the nationalisation of the coal mines and of all mineral rights. The reports differed on two issues. First the Chairman argued for 'fair and just compensation for the owners',[17] whereas the coal-miners' report laid down limits to the compensation to be paid, and three of the signatories argued against any compensation whatsoever for the owners of mineral rights. Second, the Chairman argued for state ownership on the basis of a state corporation (similar to the scheme adopted in 1946). The state would run the mines using the experience of present management (who were to be offered the opportunity of carrying on in their present jobs for at least five years) and coal-miners. Miners were to advise managers through a series of local councils. The coal-miners' report argued for a greater degree of 'workers' control', with far greater representation of workers on the industry's governing bodies.

A third report was presented by the coal-owners plus two of the government nominees. It accepted the need to nationalise mineral rights (in fact, the only major common feature of all the reports) but argued that 'the nationalisation of the coal industry in any form would be detrimental to the development of the industry and to the economic life of the country.'[18] It also suggested conciliation machinery for the industry based on a series of pit, district and national committees. These committees would be composed of representatives of mine-owners and would deal with any question of 'mutual interest'.

A fourth report was presented by Sir Arthur Duckham. He agreed with the other reports that 'the private ownership of minerals has not been in the best interests of the community',[19] and he recommended the nationalisation of mineral rights. Looking at the industry more generally, however, he argued that 'it [had] not been shown that there [was] an increased output per worker or less industrial strife when undertakings [were] owned and controlled by the state,'[20] and therefore that 'national ownership and control of collieries [did] not offer a real solution'[21] to the problems of the coal-mining industry. The solution, he suggested, lay in the unification of the industry under private ownership. Unification should take place by the amalgamation of all private companies in an area into a statutory company called a District Coal Board. The government, he suggested, should underwrite these new companies by guaranteeing a minimum dividend for shareholders but controlling excess profits by ensuring that after a certain level profits were used to reduce the price of coal.

The Sankey Commission is interesting for a number of reasons. First, it is clear that the miners saw the establishment of the Commission, in itself, as a victory. Page-Arnot, in a pamphlet compiled "for the Miners' Federation" described it as 'one of the great achievements in the history of the labour movement in any country.'[22] When one considers the outcome of the Commission it is doubtful whether such claims were really justified, yet it is clear that the miners' high hopes for the Commission were one of the main reasons they were so bitterly disappointed with the outcome. Second, the composition of the Commission was somewhat unusual. It more closely resembled a joint negotiating body or conciliation machinery in composition than a traditional Royal Commission. Third, the Commission is probably most interesting for the detail it revealed about the miners' view on nationalisation.

It might be worth-while, at this juncture, digressing a little to look into the history of the support of the miners and their unions for the

14

nationalisation of the industry. The concept of the nationalisation of the industry is, in fact, a fairly recent one; it did not really arise until the late nineteenth century. Behind it, however, lay an old 'radical' idea that land, and the minerals under the land, ought to belong to the nation and the Royal Commission of 1890 on the nationalisation of minerals provided the opportunity for the first important public discussion of the idea of the nationalisation of the coal-mining industry. The evidence presented to the Royal Commission by the coal-miners was fairly mixed; some miners' leaders argued against the nationalisation of minerals — Fenwick, a Northumberland miners' MP, said that nationalisation of minerals was 'not a principle that I am prepared to advocate in a country like ours';[23] while others argued in favour of such action — such as Samuel Woods, President of the Lancashire miners and John Wilson from Durham, Secretary of the Miners National Union. A group of miners' leaders from Scotland went even further and argued for the nationalisation not only of minerals, but also of the rest of the industry. Such views were put forward by Robert Smillie, a checkweighman at Larkhall in Lanarkshire, William Small, Secretary of the Lanarkshire miners, and James Kier Hardy, President of the Ayrshire miners. It is interesting, though, to note the reasons that were given to support the idea of the nationalisation of the industry; the central reason was the improvement of conditions in the mines, in particular safety. Thus, when asked by the Commission what advantages the nationalisation of the industry would bring, Smillie replied:

> It would lie in greater safety for the workmen. The state would look to the safety of the workmen a great deal better than is done at the present time; there would be a great deal of life saved in the mines, and a great many accidents which now occur would not occur, through better ventilation, and the mines being better looked after. The state, I believe, would have more servants to look after the safety of the workmen. It would not be a matter of expense with them, because they would know the whole expenses necessary to look after the safety of the workmen, and the working of mines would require to be paid for by the public, and it would tend to the safety of workmen and the more efficient working of the mines all through.[24]

The idea of the nationalisation of the industry was first discussed by the labour movement at the Trades Union Congress in 1892. Small proposed the resolution:

That this Congress, recognising the fact that wellnigh three quarters of a million workers are engaged in winning from the bowels of the earth produce that is a national property, is of the opinion that the enterprise should also be, like the Post Office, a state department and accordingly instructs the Parliamentary committee to prepare a bill embodying the foregoing facts and opinions.[25]

The resolution was passed unanimously. The following year Kier Hardy introduced the first nationalisation Bill into Parliament.

Despite the endorsement of the idea of the nationalisation of the industry by the general labour movement many miners' leaders were still not convinced. In 1894 at the annual conference of the Miners' Federation Greenhall (Lancashire miners) proposed the motion:

That in the opinion of this conference the best interests of the nation will be served by the nationalisation of the mines of the country.[26]

The discussion of the motion revealed that the miners' leaders were by no means as united in their support of the idea as the TUC had been two years earlier. Pickard, President of the Federation, opposed the motion because he 'didn't think that if the mines were nationalised the miners would be a penny better off.'[27] Nevertheless, the motion was carried by 158 votes to 50.

This did not, however, signal the end of the opposition within the miners' unions to the idea of nationalisation. Many of the leading positions in the Federation were held by Liberals who were far from keen on the idea. As a result, although the Federation executive was instructed to prepare a Bill for the nationalisation of the mines in 1907 no decisive action was taken for a further five years. It was not, in fact, until after the end of the first decade of the twentieth century when most of the Liberal leadership of the Federation was replaced, and in particular in 1912 when Smillie became President of the Federation, that any real action was taken.

There is also evidence that by this time there was growing rank and file support for the nationalisation of the industry. A fair proportion of the rank and file had up to this time, like the national leadership, stopped short of fully embracing socialism and the Labour Party; many had stayed loyal to the Liberal Party. In 1911 and 1912, however, there were signs that things had begun to change. In South Wales in 1911 an unofficial group of miners produced a plea for greater militancy in a pamphlet 'Miners, wake up!', and in 1912 the Unofficial Reform Committee, another product of the Welsh coal-fields, published the pamphlet 'The Miners' Next Step'. This latter pamphlet argued that the miners ought to fight for more than merely nationalisation; they ought to fight for the workers' control of the industry.

Nationalisation of the mines . . . simply makes a National Trust, with all the force of government behind it, whose one concern will be, to see that the industry will be run in such a way, as to pay the interest on the bonds, with which the coalowners are paid out, and to extract as much profit as possible, in order to relieve the taxation of other landlords and capitalists.[28]

The men themselves should

determine under what conditions and how, the work should be done. This would mean real democracy in real life, making for real manhood and womanhood. Any other form of democracy is a delusion and a snare.[29]

The pamphlet was repudiated by the Miners' Federation but it was indicative of a train of thought that was to gain ground in the coming years.

During the war the miners had seen the government intervene in the industry and although the intervention had stopped short of full control, it gave them the impetus to renew their demands for the nationalisation of the industry after the war. In 1918 at their annual conference the Miners' Federation approved a resolution renewing their faith in nationalisation.

In the opinion of this Conference, the time has arrived in the history of the coal-mining industry when it is clearly in the national interest to transfer the entire industry from private ownership and control to state ownership, with joint control and administration by the workmen and the state.[30]

and at the Sankey Commission they renewed their attack. During the hearing of evidence before the Commission the miners spelt out in more detail exactly how they saw the industry operating under nationalisation. Their plans were put to the Commission by Mr. W. Straker, Secretary of the Northumberland Miners' Association. His proposals were for the running of the industry by a new body known as the Mining Council. The council would be composed of ten members, five appointed by the Minister for Mines (two specifically representing consumer interests) and five appointed by the Miners' Federation.

Members of the Council would be appointed for five years in the first instance, although they would be eligible for re-appointment. The Mining Council would, in effect, be the governing body for the industry appointing managers and all other employees. At the district level a District Mining Council would be set up with a similar composition to its national counterpart. The District Council would exercise functions delegated to it by the National Council; it was anticipated that it would undertake much of the day-to-day supervision of the industry. It was also envisaged that there could, if the District Council thought it advisable, be pit committees, again with a similar composition to the National Council.

One needs to be careful, however, about the interpretation one puts on this plan. Certain miners' leaders clearly envisaged that it would lead to a form of "workers' control" in the industry. For example, Hodges, then Secretary of the Miners' Federation, wrote of how he saw nationalisation of the mines as the first step along the road to Guild Socialism[31] and Straker, in his evidence to the Sankey Commission, clearly saw nationalisation as a vehicle for "workers' control".

> Any administration of the mines under nationalisation must not leave the mine-worker in the position of a mere wage earner whose energies are directed by the will of another.
> He must have a share in the management of the industry in which he is engaged, and understand all about the purpose and destination of the product he is producing; he must know both the product and the commercial side of the industry. He must feel that the industry is being run by him in order to produce coal for the use of the community instead of profit for a few people. He would thus feel the responsibility which would rest upon him, as a citizen, and direct his energies for the common good.[32]

And Straker concluded his evidence by saying:

> The mere granting of the 30 per cent and the shorter hours demanded will not prevent unrest, neither will nationalisation with bureaucratic administration. Just as we are making political democracy world wide, so we must have industrial democracy in order that men may be free.[33]

However, it is clear that other miners' leaders and many miners did not share this view. They interpreted the plans for the industry much more literally; they would give the miners joint, but not sole, control of the industry. This view is supported by Barry.

Smillie, and probably most miners interpreted (the plans) literally as giving the miners, through their union, joint control with persons (unspecified) to be appointed by the Crown, who might well have been chosen from former colliery owners. At the point of production, the pit council was seen by Smillie as advisory to the pit manager who (being necessarily a qualified mining engineer) would remain personally responsible for safety measures and other technical matters.[34]

Further, in discussions on the miners' proposals, Smillie and Brace assured Lloyd George that they would not lead to full "workers' control".

In the event, the discussion over the merits and motives of nationalisation turned out to be somewhat academic, at least for the time being. The government decided to reject the reports of the miners and the Chairman and not to nationalise the industry. Instead they accepted part, although significantly not the whole, of Sir Arthur Duckham's report.

To this end in 1920 the government introduced the Coal Mines (Emergency) Act and the Mining Industry Act. The former Act was introduced in March to control the profits of the industry. The effect of the Act was to both limit and guarantee profits in a way not dissimilar to that recommended by Sir Arthur Duckham. The second piece of legislation was introduced in August 1920. The Mining Industry Act was divided into three parts. Part I deal with the administration of the industry. It established the Mines Department at the Board of Trade and set up advisory machinery for the industry. The central advisory committee was to consist of a chairman and twenty-four other persons, drawn from all sections of the industry (for example, four were to represent the coal-owners, four the coal-miners, three employers in other industries, three employees in other industries, and so on). Part II dealt more directly with the operation of the industry. Provision was made for the introduction of pit committees at every mine. The committees, with a maximum membership of ten, were to be composed of equal numbers of owners and miners. The committees were to have fairly wide jurisdiction, the scope extending from safety, health and welfare, to output and wages disputes. The pit committees were to be supported by similar bodies at district and national level, so that disputes which could not be settled at the local level could be passed up through the chain, eventually being determined at the national level. Part III of the Act was rather more general dealing with matters from the drainage of the mines to the 'social conditions of the colliery workers'.

It is clear, then, that although the government broadly accepted

Duckham's report they did not act on the whole range of his recommendations. One particularly crucial area omitted was that of amalgamations. Duckham had recommended the unification of the industry under private ownership but neither of the 1920 Acts dealt with this matter.

Not surprisingly the reaction of the miners was unfavourable. They had argued for the nationalisation of the mines and numerically at least they had gained the support of a majority of the Sankey Commission. The government, however, had disregarded not only their arguments but also those of the Chairman of the Commission, Mr. Justice Sankey. The miners determined to continue their campaign for nationalisation and appealed to the TUC for support. A special Congress was held in December to discuss this request. The miners wanted the TUC to support them by taking industrial action. Speaking at the Congress, Smillie said, 'The only thing that [will] move the government is industrial force.'[35] The TUC, however, decided to postpone a final decision on the form their support should take until the spring. In the meantime the TUC, the Labour Party and the miners agreed to jointly sponsor a massive publicity campaign under the heading 'Mines for the Nation'. Nearly a hundred demonstrations were held in cities and towns, fifteen million leaflets and tens of thousands of pamphlets were distributed and articles were printed in all trade union journals. Nevertheless, when the Congress reconvened in March 1920 it decided against 'trade union action in the form of a general strike'[36] and in favour of 'political action in the form of intensive political propaganda in preparation for a General Election.'[37] In fact, this decision meant the end of the campaign for the nationalisation of the mines. The attention of the miners and their unions once again turned to economic matters.

The miners in 1919 had gained wage increases of about 20 per cent yet it was already clear that much of the advantage gained from this advance had been lost through rising food prices. The miners, therefore, began to press for further wage increases. Following strikes in October and November of 1920 the miners and owners reached an interim agreement and began negotiations on the restructuring of wages in the industry. Some measure of agreement was reached fairly quickly; the wage levels in the industry needed to be related, to some extent, to the level of profits. The major disagreement arose over the way of looking at these profits. The miners wanted wages to be based on the national profits of the industry whereas the owners pressed for district assessments.

These discussions, however, were interrupted by a new development. The government suddenly announced that the control they had been exercising over the industry since the war would come to an end on

31 March 1921. The effect of this action, and no doubt one of the major reasons for it, was to bring the government subsidy to the industry to an end. Changes in trading conditions had resulted in a decline in the demand for coal and the government, because of its control of the industry, was faced with the prospect of having to pay a substantial subsidy. When the mine-owners heard of the government's decision they declared that the situation had significantly changed. The rates previously under discussion could no longer be considered and new district wage rates were proposed. Consequently government control of the industry ended in March 1921 with the industry in chaos. The miners refused to work on the basis of the rates proposed and went on strike for three months.

The dispute was finally resolved in July 1921. The final agreement is important because it was to set the scene for the industry's wage structure for some time to come. The agreement, known as the 'Terms of Settlement', provided a complicated system whereby wages were fixed on the lines of a basic wage plus a district percentage, set according to the profit levels in the district. The method of calculating the district percentage was as follows. The proceeds of the sale of coal in the district were ascertained and from these were deducted allowable costs. Wages were an allowable cost at a notional 1914 level and profits were an allowable cost at 17 per cent of the notional wages. If, after this exercise anything was left, then 83 per cent of it was to be added to the notional wages to make the new 'standard' wage. As a cover, it was agreed that in no district would actual wages be less than the notional wages plus 20 per cent.

Over the next few years the industry enjoyed a revival, albeit as it turned out a temporary revival, in its fortunes. The revival, like many of its predecessors, was the result of changes in the export market. Two events, coal strikes in the USA and the invasion of the Ruhr by the French, temporarily reduced the availability of coal on the world's markets and paved the way for a significant increase in the value of Britain's coal export trade. In 1923 113 million tons of coal were exported compared to 37 million tons two years earlier. Total production in Britain consequently increased from 163 million tons in 1921 to 276 million tons in 1923 and profits in 1923 at over £27 million were more than double their 1922 level.

The miners' reaction to this increasing prosperity was to ask for a review of the 1921 agreement. They gave notice of their intention to terminate this agreement and the notice expired on 17 April 1924. Afterwards working was continued on the basis of a temporary agreement and the government set up a Court of Inquiry under Lord Buckmaster[38] to investigate the dispute. The inquiry did not produce any definite proposals for settling the dispute but it did lay down some of the guidelines. It recognised, for example, that the earnings of day-wage workers were substantially less than their pre-war equivalents and suggested that the provision of an adequate minimum wage should have precedence over the distribution of profits. Following the publication of the report the two sides entered into new discussions and reached an agreement whereby the minimum addition to the standard wage would be increased from 20 per cent to 33½ per cent, and the percentage of surplus profits allocated to wages would be 87 per cent.

The boom of the early 1920s, however, was soon to pass. Once again, exports were at the heart of the matter. The conditions which had given Britain a temporary advantage vanished and British coal exports declined from 113 million tons in 1923 to 71 million tons in 1925. Total production similarly fell from 276 million tons in 1923 to 243 million tons in 1925, and the economic position of the industry deteriorated so that a profit of £27 million became a loss of £2 million (first six months of 1925). The coal-owners were just as quick as the miners to make use of changed conditions and now they demanded a revision of the wages agreement.

They argued that the industry's problems were largely the result of excessive costs, and that these in turn were the result of high wages. The mine owners suggested, therefore, a revision of the agreement which would have effectively led to a reduction in wage levels. Probably the most important step in this direction was the abandoning of the minimum wage provisions. The mine-workers rejected these suggestions and argued that most of the industry's problems arose from poor organisation; the miners therefore should not be asked to bear the brunt of any sacrifices.

When negotiations between the two sides broke down the government appointed another Court of Inquiry, this time under the chairmanship of Mr. H. P. Macmillan, to look into the dispute. The report of the inquiry[39] argued that the industry's problems were 'to a large extent the creation of neither party to the dispute'.[40] They were the result of 'national and international disturbances and dislocations which are the inevitable sequel of the economic upheaval due to the war'.[41] It concluded that the miners were justified in demanding a retention of the minimum wage, although it was unwilling to say what the minimum wage should be. It also suggested that there was 'considerable room for improving the efficiency of the industry as a whole'[42] and thus reducing costs.

In an attempt to avoid a stoppage of work the government proposed to give a subsidy to the industry to bridge the gap between the wages due under the 1924 Settlement and those which would have been due under the employers' proposals. This would provide a breathing space during which another, fuller inquiry would be carried out into the industry.

The government paid the subsidy and avoided an immediate stoppage of work. The subsidy, however, was £23 million, not the £10 million they estimated. The inquiry was carried out by a Royal Commission, with Sir H. L. Samuel as chairman. The Commission was appointed on 5 September 1925 but did not present its report until 6 March 1926.

The Commission produced a far more detailed and discursive report[43] than the previous inquiries. It highlighted the cost position of the industry as the major problem.

> The dominant fact is that, in the last quarter of 1925, if the subsidy be excluded, 73 per cent of the coal was produced at a loss.[44]

The answer, they stressed, was not a continuation of the present subsidy.

> We express no opinion whether the grant of a subsidy last July was unavoidable or not, but we think its continuance indefensible. The subsidy should stop at the end of its authorised term, and should never be repeated.[45]

The answer rather lay in the attack on the costs of the industry.

> The gap between proceeds and costs in mining can in the near future be filled only in two ways: by a sudden contraction in the industry to much smaller dimensions and a rise in prices, or by an immediate lowering of cost of production. Some contraction of the industry is probably inevitable, and, in so far as it can be limited to the closing of definitely inefficient pits, it is desirable in spite of the distress that it must cause. The scale of contraction indicated by the figure of losses just given is of an altogether different order from this; it means not the disappearance of the inefficient, but the collapse of an industry. The second way of filling the gap cannot be avoided. We come reluctantly but unhesitatingly to the conclusion that the cost of production, with the present hours and wages, are greater than the industry can bear.[46]

The report rejected the idea that much could be done by increasing hours. This would make the working day of the British miner longer than any other in Europe except Poland (Upper Silesia) and would probably lead to increasing absenteeism and 'slackness of work'. The answer, then, lay in a reduction in wages. This must mean giving up the minimum wage established in 1924.

> We can see no escape from giving up this minimum of 1924. That does not simply mean a return to the minimum of 1921; i.e. a uniform reduction of 10 per cent on the present minimum percentage. In some districts less may be needed, in others more. It will be for the mine-owners and the miners to carry out by negotiation the necessary downward revision of district minima.[47]

The report in many ways then clearly favoured the owners' or employers' point of view rather than that put forward by the miners. It recommended a reduction in wages as the main way of overcoming the industry's problems. Yet it did not go all the way with the employers. They also wanted an increase in working hours, which the Commission rejected. In addition, although the report recommended the abolition of the minimum wage, it argued that the wages of the lowest paid men should be safeguarded by continuing the system of subsistence payments. It also recommended that other ways of reducing costs be looked at, such as improving organisational efficiency (although it rejected the idea of nationalisation as a way of solving the problems).

The report concluded with an exhortation for greater co-operation between mine-owners and mine-workers. If this could be achieved then the future was secure.

> The future depends primarily upon the leadership, and the general level of opinion, among the mine-owners and the miners of Great Britain. In laying down our charge, we would express our firm conviction, that if the present difficulties be wisely handled, if the grievances of the one side and of the other be remedied, and a better spirit prevail in consequence between them, the mining industry, with the aid of science, will certaintly recover, and even surpass, its former prosperity. It will again become a source of great economic strength to the nation.[48]

The government, in effect, accepted the report but did little about it. The result was that instead of the cooperation pleaded for by the Commission there ensued one of the most celebrated and destructive conflicts in British labour history. Although the General Strike itself lasted only a little over a week (3-12 May) the miners stayed on strike for ten months. Bolstered by Churchill's comment that 'it's the length of the stoppage that matters',[49] and the knowledge that the strike had been estimated to be costing the country £3 million a day, the miners held out on their own. In the end, however, they were forced to concede on almost every issue. In the course of the stoppage an Act of Parliament was passed permitting a return to the 'Eight-Hour Day' and when work was eventually resumed in November 1926 it was on the basis of a reduced minimum wage and district assessments. The significance of the General Strike has been debated at length, yet as far as the coal-mining industry is concerned Court's comment seems particularly apt.

The main importance of the dispute of 1925-26, therefore, lies in this, that the solution adopted for the slumping sales and profits of the early 'twenties was certainly not that of the miners, who had taken up an impractically rigid and long-term position in favour of any other costs being reduced except labour costs; and was not even that believed to have preferred by the government of the day, who inclined towards the middle way of the Samuel Commission, without nourishing fervour enough to carry it into law; but was entirely that suggested by regional competition within the industry. It took the form of an attack upon labour costs, to the exclusion of other costs, by the most simple and direct way, the alteration of wages and hours. The way in which this was done embittered the relations of management and men for the rest of the inter-war years and made extremely difficult the introduction of other methods of reducing costs which were essential; for those required for their perfect success the intelligent and willing cooperation of the miners.

After the General Strike

If it was hoped that the experiences of 1925 and 1926, however painful, might at least lead to a more stable, profitable and efficient industry then these hopes were surely dashed. The period from the General Strike to the beginning of the Second World War was an extremely poor one for the industry, certainly in economic terms.
Production fell fairly consistently from 1927 to 1933 (see Table 3). By 1933 only 191 million tons of coal were being mined compared to a pre-General Strike total of over 221 million tons (nearly 244 million tons in 1924). Although production rose after 1933 it did not reach the pre-General Strike total again until 1937. Profits suffered a similar fate. In 1927 and 1928 a loss was made and although the position improved somewhat after 1933 (see Table 3) profits never rose above 1s. a ton until 1937 (the 1933 profits might be compared to the high point of 2s. 2d. a ton in 1923). Further these figures refer to profits per ton mined. Total profits clearly fell by a much larger

percentage, especially during the late 1920s and early 1930s. Thus, for example, between 1930 and 1934 profits in the industry were running at only about £3 million a year compared to the £27 million of 1923.

The decline in coal output and the industry's profitability is clearly related to a number of factors but two particularly important ones might be stressed. First, home consumption of coal expanded very little in the inter-war years. This was partly a result of the general economic depression but also partly a result of the increasing use of substitutes for coal (especially oil) and greater efficiency in fuel burning. Thus, home consumption at 185 million tons in 1938 was virtually the same as it had been immediately prior to the First World War.

A second, and probably more important, factor was the decline in exports. By 1932 less than 40 million tons of coal were being exported from Britain each year (compared to 113 million tons in 1923) and in 1938 the total was still only 46 million tons. The decline in exports was, itself, related to two factors.

First, world demand for coal grew very little after 1913. In the years before 1913 world demand had increased on average by about 4 per cent a year but after the First World War there was virtually no growth in demand at all (on average only 0.2 per cent a year). Coal, world-wide, as in Britain, was suffering from the challenge of oil, and the improved coal-burning techniques.

Second, Britain's share of the world coal market had started to decline. Britain had to face strong European competition, particularly after the development of the German and Polish coalfields in the 1920s, but because of the industry's poor organisation (the industry was still dominated by a large number of small pits),[51] the lack of investment[52] and the high exchange rate for the pound (particularly after the return to the Gold Standard in 1925), she was unable to meet this challenge.

Attempts to improve the structure and organisation of the industry

The government reacted to these problems in a very tentative fashion. They tried various measures to improve the organisation of the industry, in line with the recommendations of the Samuel Commission,[53] but without much support from the coal-owners and, in the end, without a great deal of success.

The government's first move in this direction came in 1926 with the Mining Industry Act of that year. The Act dealt with a variety of issues from recruitment to welfare, to the establishment of profit-sharing schemes. Undoubtedly the most important part of the

26

Act, however, was Part I which dealt with the reorganisation of the industry. Under Part I facilities were provided for the preparation of amalgamation schemes. Such schemes could be prepared even though one of the owners was 'unwilling to agree to amalgamate or to agree to the proposed terms of the amalgamation.'[54] Schemes, voluntary or not, could be referred to the Railway and Canal Commission for consideration and on their recommendation be made binding.

The Act had some, but relatively little effect. A Board of Trade report[55] in 1928 showed that apart from 1 major amalgamation scheme covering 61 pits, only 13 other schemes, covering 130 pits, had been completed. The chief problem was that although the Act removed some of the impediments to amalgamation it left the initiative to the industry itself. The industry, despite protestations to the contrary, did not seem to have the will to reorganise on this basis. The industry, in fact, seemed rather more interested in devices of a different kind. Taking their cue from another government report issued during the General Strike,[56] a number of owners made attempts during the late 1920s to introduce cooperative selling and marketing schemes. Undoubtedly the most successful of these was the 'Five Countries' scheme. Started in 1928, it originally covered the Yorkshire, Nottinghamshire and Derby coalfields but it quickly spread into Warwickshire, Cannock Chase, North Staffordshire, Leicestershire, Lancashire and Cheshire. The scheme operated on three lines. First, it aimed to regulate the output of the coal-fields in conformity with market requirements. Second, it aimed, with a policy of subsidies, to help extend the export trade. Third, it organised the marketing of coal both at home and abroad.

In 1930 the government introduced further legislation in an attempt to deal with the industry's problems. The Coal Mines Act of that year was a major piece of legislation. It was divided into four main parts; Part I dealt with production, supply and marketing; Part II with the organisation of the industry; Part III with hours of work; and Part IV with the setting up of the Coal Mines National Industrial Board.

Part I of the Act really built on the industry's earlier voluntary efforts in this area. It attempted to deal with the industry's problems by regulating the production and supply of coal. Under these provisions the industry was to be subject to restrictions on the amount of coal it could sell and, to a certain extent, the price of the coal it sold. The restrictions were to be imposed by one central council and a number of district councils. These councils would have the responsibility for setting quotas and prices to be charged.

Part II of the Act established a Coal Mines Reorganisation Commission, consisting of five Commissioners appointed by the

Board of Trade. The Commission was instructed to assist in the further reorganisation of the industry by promoting amalgamations 'where such amalgamations appear to the Commission to be in the national interest'.[57] Schemes for amalgamation proposed by the Commission were to be dealt with under the provision of the 1926 Mining Industry Act.

Part III of the Act reduced the hours of work in the coal mines by half an hour a day. The 'Eight-Hour Day' which had been introduced during the General Strike of 1926 was, through this Act, reduced to a 'Seven and a Half-Hour Day'.

Part IV of the Act established the Coal Mines National Industrial Board. The National Board was to be composed of seventeen members of whom six were to be representatives of the coal-owners, six representatives of the coal-miners, and the other five were to be representatives of various national employer and employee organisations. The National Board had the function of overseeing agreements between owners and workers and conciliating on disputes in the industry.

The Act, despite its wide coverage, had very limited success. Part I of the Act was in essence a temporary expedient designed to spread and maintain employment in the industry. It had only limited success for, although earlier the industry had itself made moves towards centralised marketing and the establishment of quotas, this did not prevent many owners from trying to find ways round the Act's provisions. Among the methods tried were:

> the setting up of bogus agencies and subsidiary companies to facilitate sales below the minimum, bonus payments to customers, systematic over-weight and allowances for short weight, contra purchases, juggling with transport charges and misdescription of coals and their destination.[58]

Part II of the Act was designed to give more teeth to the amalgamation proposals of the 1926 Mining Industry Act. It was fairly ineffective, however, for three reasons. First, Part I of the Act, by laying down quotas for individual mines, encouraged the retention of marginal and unprofitable mines. In this way Part I of the Act effectively discouraged amalgamations. Thus, a PEP report on the industry said:

> Parts I and II of the 1930 Act are essentially mutually conflicting. The quota scheme embodied in Part I, however expedient it may have been as an instrument of national policy at a time when unemployment was rapidly increasing, does not

face the real problems of the industry, and does not represent a long-term policy.[59]

Second, the Act laid down what in the end proved to be almost prohibitive restrictions on the work of the Commission. It stated that the Railway and Canal Commission should not confirm any scheme presented to it by the Coal Mines Reorganisation Commission

unless satisfied —
  (i)   that it would be in the national interest to do so, and
  (ii) in the case of an amalgamation scheme that the scheme
        (a) will result in lowering the cost of production or disposal of coal, and
        (b) will not be financially injurious to any undertakings proposed to be amalgamated, unless the scheme contains provisions for the purchase, at a price to be fixed in default of agreement by arbitration, of any such undertaking; and
  (iii) that the terms of the scheme are fair and equitable to all persons affected thereby.[60]

The interpretation of this clause (inserted as an amendment by the House of Lords) effectively ruled out most schemes. In particular, the requirement that the scheme should not be financially injurious to any undertaking was strictly interpreted by the courts and extremely difficult to comply with.

Third, Part II of the Act was effectively resisted by the owners who were willing to use all possible methods to evade its provisions. Although some owners would have welcomed a programme of amalgamations and a thorough overhaul of the industry's structure, the majority disliked the Act and tried to make it unworkable. They regarded the Reorganisation Commission 'merely as a temporary inconvenience which need not be taken too seriously',[61] and as Wilson asserts, the Commission never had 'the powers necessary to force amalgamations on an unwilling and recalcitrant industry'.[62]

The two other parts of the Act suffered a similar fate. The new provisions on hours were diluted when it was agreed that the seven and a half-hour day could be averaged out over a two-week period and the National Board proved to be an ineffective institution. The coal-owners, because they wanted to preserve the idea of district bargains and negotiations, refused to nominate members to the Board.

In the mid-1930s the government took two important steps concerning matters covered by the 1930 Act. First, in 1936 it effectively blocked any further work on Part II of the 1930 Act by asking the Reorganisation Commission not to proceed with any more amalgamation schemes until a full inquiry into the working of this section had been instituted. Second, in 1935 the government started moves to strengthen Part I of the 1930 Act. As one of the terms of the settlement of a wages dispute the coal-owners agreed to reinforce the marketing and selling controls of the 1930 Act, and as a result of this agreement three types of scheme were introduced. In one type of scheme a central selling organisation was established to market coal for the district. In another, although individual coal-owners continued to market their own coal, sales were closely controlled (not only in respect of price, but also in respect of quantity and distribution of coal) by a central committee. In the third, selling was undertaken by a group of owners, the policy of these groups being coordinated by an Executive Board for the district.

The government's final attempt before the war to improve the structure of the industry came in 1938 with the Coal Act of that year. The Act intervened in two important areas. First, the Act revised the position with regard to the reorganisation of the industry. The duties of the Coal Mines Reorganisation Commission were to be taken over by a new body, the Coal Commission. This body was to be given increased powers to investigate the desirability of amalgamations, and if on investigation it felt that although amalgamation was desirable it did not possess sufficient powers to carry the amalgamation through then it could apply to the Board of Trade for an extension of its powers.

The mine-owners objected strongly to this part of the Act. They objected on two grounds.

> The first is that compulsory amalgamation is "bureaucratic" and unnecessary, since voluntary amalgamations are proceeding with all desirable speed. The second is that the "safeguards" incorporated in the compulsory amalgamation procedure of the 1930 Act have been removed.[63]

The mine-owners, however, found little support for their arguments. In the past they had, by most measures, shown reluctance and an inability to reorganise the industry themselves and few people therefore opposed government intervention. Thus *The Economist*, commenting on the mine-owners' argument said:

> The practice of compulsory amalgamation was settled in 1930 by a Parliament composed of a Labour and Liberal House of Commons and a Conservative House of Lords. The present Bill does no more than make a second attempt to implement what was

the manifest intention of Parliament eight years ago – surely not an excessively rash rate of progression. As for the so-called "safeguards", they have merely been amended to prevent them from being, as they have since 1930, absolute barriers to any action being taken. When the mine-owners in their published statement by implication bless the 1930 Act with its "safeguards", they mean that they support the principle of compulsory amalgamation provided that, in practice, no compulsory amalgamations take place. It is an attitude to which, on the lips of less high-minded men, harsh descriptions might be applied.[64]

In the event this part of the Act did not have a great deal of impact, partly because its provisions were still not strong enough and partly because of the intervention of the war. The Act had rather more impact, however, in another area.

The 1938 Act also provided for the nationalisation of coal rights. The nationalisation of coal rights had been recommended as far back as 1919 by the Sankey Commission. It was, in fact, one of the few matters on which all members of the Commission had agreed. The nationalisation of coal rights had, however, been strenuously opposed by many, the length of time it had taken to bring the measure about being an indication of the strength of the opposition. Nationalisation of coal rights was, of course, a far cry from the nationalisation of the coal mines, yet it was clearly of considerable benefit to the industry. The disadvantages of the private ownership of coal rights are well summed up in the following quotation.

The disadvantages are various in character, some are technical. Thus, the layout of a minefield has to be governed largely by surface boundaries and surface rights; a mine-owner sinking a shaft has to obtain leases from a multiplicity of landowners (the average number is five), and any of these may, if he chooses, cause endless trouble and expense by refusing to lease or demanding unreasonable or exorbitant terms; barriers of coal have to be left between the property of one royalty owner and that of another. Other disadvantages are financial. The charges levied for royalties and for wayleaves, both surface and underground, are very heavy. The average royalty charge is 5.78d. per ton (1935); but the highest recorded is 1s. 9d. per ton (in South Wales). The total sum paid out in 1935 was £4,806,139. Even between whole districts there are very wide variations. Thus the average in South Wales is 8.48d. per ton and in South Derbyshire 3.82d. per ton of coal commercially disposable. Broadly speaking, the charges are heaviest in the least profitable districts, and especially in the export districts. Finally, the private ownership of royalties has a serious political disadvantage; the workers are embittered by the knowledge that very large sums are paid every year to fortunate individuals who contribute nothing to the carrying on of the industry.[65]

The Effect on the Miners

The owners and the employers were not the only ones to feel the effect of the industry's poor performance between 1926 and 1939. The miners bore as many, if not more, of the consequences and were significantly affected in two crucial areas, employment and wages. The reduction in the demand for coal meant a reduction in the number of men employed in the industry (see Table 4) and a significant increase in unemployment (see Table 4). Thus employment in the industry declined from its pre-General Strike heights of over 1,200,000 to about 800,000 by the beginning of the Second World War. Unemployment was consistently high (above 15 per cent) through the period, reaching a record 34 per cent in 1932. The reduction in the profits of the owners meant increasing pressure to reduce wages. The reductions (see Table 5) took place over a period of years and earnings reached a post-war low in 1932. Even taking the general fall in wage levels into account the miners were still considerably worse off in 1932 than they had been before the General Strike.

The coal-miners' reaction was typically stubborn. The period was one of tremendous industrial conflict. The bitterness that resulted from the General Strike, allied to the reduction of wages, and the frequent colliery disasters, created an atmosphere of distrust and hatred. The number and extent of stoppages of work during this period is by now well known and well documented. Table 6 shows that between 1927 and 1939 the industry, which covered little more than 5 per cent of the employed population, was frequently responsible for as many as 40 to 50 per cent of man days lost through stoppages in the country as a whole.

Throughout much of this period, though, the miners and their trade unions were on the defensive. The trade unions were slow to recover from the effects of the General Strike and were not helped by their comparatively weak bargaining position in the late 1920s and early 1930s. Further, they had to contend with opposition from rival workers' organisations.

The most serious such opposition came from the 'Non-Political' or 'Spencer' Unions. These unions, known as Spencer Unions after their President George Spencer, were established during the General Strike in Nottingham. They were strongly supported by the Nottinghamshire coal-owners who gave them exclusive bargaining rights and certain financial help. In December 1926, for example, the Nottinghamshire coal-owners offered financial support for the non-political unions' pension scheme. Such support brought charges that the unions were merely 'puppets of the owners'. Spencer denied this claiming that the owners' motives in offering financial support were merely the wish 'to see a happy, contented, united, prosperous family of miners'.[66]

The new unions had to face considerable opposition not only from the Miners' Federation but also from the rest of the labour movement. The TUC placed their full support behind the efforts to destroy the Spencer Unions because they saw in them 'a challenge to the whole organised working-class movement'.[67] They felt that the Spencer Unions had to be dealt with quickly and decisively if 'employers in other industries [were] not to be encouraged to repeat similar tactics on a wider scale'.[68]

Yet, no doubt aided by the economic conditions of the time, the Spencer Unions were far from destroyed. They spread out from their Nottinghamshire base to other parts of the Midlands and South Wales. It was not until 1936 that the Federation managed to eliminate this source of opposition. It was finally brought to an end following a protracted industrial dispute late in that year.

Following a dispute, first a Haworth Colliery over the ill-treatment of two boys, and later at Cater Colliery over the dismissal of two men for taking 'snap' (lunch) time without authority, the Miners' Federation stepped into try to negotiate on the miners' behalf. Despite a ballot held at the collieries which showed overwhelming support for the Miners' Federation, the mine-owners refused to negotiate with the Federation and insisted that the men should be represented by the Spencer Unions. The dispute escalated when the Miners' Federation gained the agreement of their members for a national strike in support of their case. At this point political intervention helped to resolve the dispute and finally end the conflict between the two unions. The dispute at the colliery was settled and plans were made for the amalgamation of the Spencer and Federation unions. A new union, the Nottinghamshire and District Miners' Federation Union was to be set up. The new union would join the Miners' Federation of Great Britain. Its officers were to be Spencer (President) and Cooper (Treasurer) from the Spencer Unions and Coleman (Secretary) from the Federation-supported union. Provision was made for Spencer to represent the new union on the Federation executive for five years, after which time the representation would be decided by election.

The weakness of the miners' unions during this period is probably best illustrated by reference to the debate over the miners' hours of work. Legislation had increased the miners' working day to 'eight hours' during the General Strike. When the Labour Government gained power in 1929 the miners had high hopes of a reduction in their working day, at least to 'seven hours'. In fact the government only reduced the length of the working day by half an hour, and even part of this benefit was eroded by a decision of the House of Lords that the seven and a half-hour day could be averaged out over two weeks. In 1931 the miners saw yet another opportunity to reduce the length of their working day. Current legislation on the length of the working day was due to expire and unless new legislation was introduced the working day would automatically be reduced to seven hours. In the event, and despite the bitter opposition of the miners' unions, the government introduced a bill in July 1931 which 'temporarily' continued the 'seven and a half-hour day'. A year later the temporary provisions of 1931 were extended by the government, again despite bitter opposition from the miners and their unions, for a further five years.

Towards the middle of the 1930s, however, conditions in the industry did begin to improve a little, and the miners' unions were able to take full advantage of this change in fortune. In 1935 they mounted what was to be one of their most successful inter-war campaigns.

Their campaign had two main platforms. First they argued for a two shillings a day increase in miners' wages across the board. Second, they argued for the establishment of national negotiating machinery for the industry. The tactics used by the miners during the campaign are particularly interesting. On the one hand, they argued their case publicly with a great deal of skill. They forcefully attacked the structure of the industry, arguing that this was the main reason for its poor profitability and offered to submit their claim to arbitration. The offer to submit the claim to arbitration was particularly useful; it clearly had a valuable public relations effect, in emphasising the 'reasonableness' of the miners' position, yet if the owners had accepted arbitration on a national claim they would in fact have been going a long way towards accepting the wisdom of national negotiating machinery, one of the miners' main demands. At the same time the miners gave an impressive show of strength of the more traditional kind. A national ballot of members was held on the question:

Are you in favour of authorising the executive committee to press the claim for an advance of wages of 2s. per shift for adults and 1s. per shift for youths even to the extent of tendering your notice, if necessary, to enforce the claim.[69]

The result of the ballot showed 409,351 in favour and only 29,215 against. On receipt of this result the executive called a special miners' conference in December and it was agreed to hold a national miners' strike, starting 27 January 1936.

The key to the miners' victory, however, lay in yet another factor. Once the miners threatened to strike then their wage claim was supported by an impressive range of coal-users. The coal-owners had argued that wage increases could not be granted because the market would not stand the necessary consequent increase in the price of coal. The support given to the miners' case by coal-users, even if it meant paying higher prices, significantly tipped the balance of the scales in the miners' direction. Although the miners in the end did not receive the increase that they had been demanding, they did gain district increases averaging 10 per cent or 9d. a day and the establishment of a body known as the National Standing Consultative Committee. The Committee, composed of representatives from both sides of the industry, was set up

> for the consideration of all questions of common interest and general application to the industry, not excluding general principles applicable to the determination of wages by national agreement.[70]

This body was, except in name, the national negotiating machinery the miners had been asking for.

Conclusion

Between 1913 and 1939 the position of the coal-mining industry world-wide altered quite dramatically. Coal was being overtaken by new sources of fuel. Oil in particular gained considerably in importance; for example, in 1913 only 3 per cent of the world's mercantile fleet had been fitted with oil-burning appliances, whereas by 1939 the figure had soared to over 53 per cent. Where coal was still used it was being used much more efficiently; new techniques had been designed to help industry gain more power from the same amount of coal.

Britain's coal-mining industry had changed as much if not more than the world-wide pattern. By 1939 it was producing about 236 million tons of coal a year, compared to over 287 million tons in 1913; the industry employed about 800,000 men in 1939 compared to over 1.1 million in 1913; and in 1939 only 46 million tons of coal was exported compared to about twice as much in 1913.

It would be a mistake, though, to paint a picture of a steady uniform decline in the industry's fortunes in Britain between 1913 and 1939. The history of those years was, in fact, much more chequered than that. From a high point in 1913, the industry was afforded temporary and perhaps artificial importance during the First World War. After the First World War the industry again benefited from unusual factors which led to an increase in the overseas demand for British coal. The industry's fortunes fluctuated for a while in the early 1920s and it was not until 1925-26 that the bubble effectively burst. The coal-mining industry after the General Strike suffered not only the effects of the Depression but also the effects of deficiencies in its own organisation and competitive position. There was a minor rally after 1934 but the industry had clearly fallen from the heights of 1913 and was never to regain them.

Government actions during this period are interesting. Between 1914 and 1918 the governments were clearly concerned to maintain coal supplies essential for the war effort. They were unwilling, though, to take any more action than was absolutely necessary. At first they tried to deal with the situation through the Coal Mining Organisation Committee and it was only the threat of industrial action that forced them to take more decisive steps. Even then they stopped short of assuming complete control of the industry.

At the end of the war the government seemed undecided on future policy. Between 1918 and 1926 it set up four major inquiries into the coal-mining industry; yet probably its most significant action was to divest itself on the responsibilities it had retained from the wartime period. It did little to implement the major recommendations of the inquiries. No action was taken, for example, over the nationalisation of the mines or even mineral rights (mineral rights were not nationalised until 1938).

After the General Strike governments seemed to have clearer objectives. They were obviously keen to see the industry reorganised and promote the amalgamations suggested by the Samuel Commission. The 1926, 1930 and 1938 Acts, however, were not strong enough to produce the desired results and the government was effectively sidetracked into centering much of its attention on cooperative marketing and selling schemes.

Throughout the 1913 and 1939 period the actions of the various governments were essentially *ad hoc* and tentative. They were *ad hoc* because they were frequently determined by the needs of a particular crisis rather than to have been based on long-term planning. They were tentative because even when they knew what course of action they would like to pursue governments rarely acted decisively to enforce their views.

The owners/employers, on the other hand, were rather more consistent in their attitude. It was dominated by two considerations; the wish to avoid government 'interference' and the belief that the level of wages held the key to the competitiveness of the industry. *The Economist's* comment that 'It would be difficult to find fiercer individualists anywhere in the world'[71] could well be supported. The owners objected strongly to the controls imposed during the First World War and announced their total opposition to any further government intervention in the industry, except for the nationalisation of mineral rights. When the government attempted to intervene to force the industry to reorganise the owners typically reacted by publicly proclaiming that such intervention was unnecessary and privately using all possible methods to thwart the government's actions.

The owners' reaction to changes in economic conditions was predictable. When economic conditions were favourable they were willing to concede wage increases; although token resistance was an essential part of the game. When economic conditions were unfavourable then wages were the first item to come under attack. The owners clearly saw labour costs as the balancing factor in their economic equation.

Nationalisation of the industry was not discussed by the miners until near the end of the nineteenth century. Initially a number of miners and their leaders rejected the idea but by the beginning of the First World War most had been persuaded to accept it. The miners in their proposals to the Sankey Commission indicated that they felt nationalisation should be accompanied by a degree of 'workers' control' although there were different views about exactly what degree of 'workers' control' was desirable. The government's rejection of the idea of nationalisation following the publication of the Sankey Commission Report, however, effectively signalled the end of the miners' campaign for some time and their attention quickly reverted to economic questions.

Their attitude towards economic questions was in some ways contradictory. On a number of occasions they argued that most of the industry's problems were the result of its poor structure and organisation. Thus, they often opposed attempts to reduce wages in order to improve the industry's competitive position (as in 1926). Yet at other times they seemed to accept that wage levels would be tied to the market position of the industry. The agreements negotiated in the 1920s are probably the best example of this kind of view. To a certain extent, of course, all union/management relations are governed by the economic position of the industry or industries concerned. In the coal-mining industry, however, the relationship between wages and the economic position of the industry was far closer and far more explicit.

37

NOTES

1. W. H. Court, *Coal,* Longmans, Green & Co., London, 1951, p.6.
2. *The Times,* 4 June 1913, p.17.
3. R. W. Dron, *The Economics of Coal Mining,* Arnold, London, 1928, p.1.
4. Ibid., p.1.
5. Terms of Reference of Departmental Committee, Coal Mining Organisation Committee.
6. Coal Mining Organisation Committee, *First Report,* Cd. 7939, H M S O London, 1915.
7. Coal Mining Organisation Committee, *Second Report,* Cd. 8147, H M S O, London, 1916.
8. Coal Mining Organisation Committee, *Third Report,* Cd. 8345, H M S O, London, 1916.
9. See R. Page Arnot, *The Miners : Years of Struggle,* Allen & Unwin, London, 1953, pp. 164-70 for detailed description of events in South Wales during 1915 and 1916.
10. Quoted at p. 168, ibid.
11. Defence of the Realm Consolidation Act, 1914.
12. Figures quoted by R. Page Arnot, *Further Facts from the Coal Commission,* Labour Research Department, London, 1920, p.8.
13. Coal Industry Commission, *Interim Report* (Chairman Sankey), Cmd. 359, H M S O, London, 1919.
14. Quoted by R. Page Arnot, *Further Facts from the Coal Commission,* Labour Research Department, London, 1920, p.8.
15. Ibid., p.8.
16. Coal Industry Commission, *Final Report* (Chairman Sankey), Cmd. 361, H M S O, London, 1919.
17. Ibid., p.5.
18. Ibid., p.35.
19. Ibid., p.49.
20. Ibid., p.50.
21. Ibid., p.53.
22. R. Page Arnot, *Further Facts From the Coal Commission,* Labour Research Department, London, 1920, p.11.
23. E. E. Barry, *Nationalisation in British Politics: The Historical Background,* Jonathan Cape, London, 1965, p.111.
24. Ibid., p.118.
25. Ibid., p.120.
26. Ibid., p.122.
27. Quoted by R. Page Arnot, *The Miners: Years of Struggle,* Allen & Unwin, London, 1953, p.35.
28. Unofficial Reform Committee, *The Miners' Next Step: Being a Scheme for the Reorganisation of the Federation,* Pluto Press, 1973 (reprint).
29. Ibid., p.32.
30. R. Page Arnot, *The Miners: Years of Struggle,* Allen & Unwin, London, 1953, p.128.
31. F. Hodges, *Nationalisation of the Mines,* Leonard Parsons, London, 1920.
32. Quoted by R. Page Arnot, *Further Facts from the Coal Commission,* Labour Research Department, London, 1920, p.37.
33. Ibid., p.37.
34. E. E. Barry, op. cit., p.241.

35. Ibid., p.242.
36. Ibid., p.243.
37. Ibid., p.243.
38. Ministry of Labour, *Report of a Court of Inquiry Concerning the Wages Position in the Coal Mining Industry* (Chairman Buckmaster), Cmd.2129 H M S O, London, 1924.
39. Ministry of Labour, *Report of a Court of Inquiry Concerning the Coal Mining Industry Dispute 1925* (Chairman MacMillan), Cmd. 2478, H M S O, London, 1925.
40. Ibid., p.20.
41. Ibid., p.20.
42. Ibid., p.19.
43. Coal Industry Commission, *Report* (Chairman Samuel), Cmd. 2600, H M S O, London, 1926.
44. Ibid., p.235.
45. Ibid., p.235.
46. Ibid., p.227.
47. Ibid., p.228.
48. Ibid., p.237.
49. Quoted in support of the view that a miners' victory was inevitable by A. J. Cook, *The Coal Shortage: Why the Miners Will Win,* Labour Research Department, London, 1926.
50. W. H. Court, op. cit., p.13.
51. By 1939 there were still over 2,000 mines operating in Britain.
52. For a discussion of the problems this caused see Ministry of Fuel and Power, *Report of the Technical Advisory Committee on Coal Mining* (Chairman Reid), Cmd. 6610, H M S O, London, 1945
53. Op. cit.
54. Mining Industry Act, 1926, Clause 1 (2).
55. Board of Trade, *Report on the Working of Part I of the Mining Industry Act 1926,* Cmd. 3214, H M S O, London, 1928.
56. Board of Trade, *Reports of Departmental Committee on Co-operative Selling in the Coal Mining Industry* (Chairman Lewis), Cmd. 2770, H M S O, London, 1926.
57. Coal Mines Act, 1930, Clause 12 (1).
58. Times Trade Supplement, 18 August 1933. Quoted in I. Thomas, *Coal in the New Era,* Putnam, London, 1934, pp.190-91.
59. Political and Economic Planning, *Report on the British Coal Industry,* P E P, London, 1936, p.5.
60. Coal Mines Act, 1930, Claus 13 (2).
61. Coal Mines Reorganisation Commission, *Reports on the Working of the 1930 Act,* quoted by I. Thomas, op. cit., p.192.
62. H. Wilson, *New Deal for Coal,* Cole, London, 1945, pp.22-3.
63. *The Economist,* 5 February 1938, p.277.
64. *The Economist,* 5 February 1938, p. 278.
65. *The Economist,* 24 October 1936, pp.153-4.
66. *The Times,* 20 December 1926, p.12.
67. *The Economist,* 5 May 1928, p.922.
68. Ibid.
69. *The Times,* 12 November 1935, p.14.
70. *The Economist,* 11 February 1936, pp.227-8.
71. *The Economist,* 11 April 1942, p.490.

3. WARTIME PROBLEMS

The government was clearly concerned to ensure that the war effort would not be hampered by the lack of coal during the Second World War as it had been during the First. At the outbreak of war the government took a range of powers designed to secure the position of coal supplies. The Schedule of Reserved Occupations,[1] the Armed Forces (Conditions of Service) Act[2] and the National Service (Armed Forces) Act[3] were used to protect manpower in the industry; the government took over control of the marketing of coal by appointing Coal Supply Officers in all regions to oversee the industry's centralised selling schemes; and rationing was introduced for domestic consumers through the Fuel and Lighting Order[4] (domestic and small industrial consumers were required to restrict their purchases of coal to 75 per cent of what they had been in the corresponding quarter of the previous year). The first part of the war, however, was an unstable period for the coal-mining industry. Between the outbreak of war and the middle of 1941 demand for coal fluctuated violently, and government policy followed in a similar fashion.

Fluctuating Fortunes

For the first few months of the war there was, in fact, no real shortage of coal. In November 1939, as a result, the government effectively abandoned its scheme for rationing domestic users. Early in 1940 the position did worsen somewhat. There were frequent reports of coal shortages, particularly in the South East.

> The shortage of house coal has left many families without adequate heating when it was most badly needed, and this may have contributed to the widespread incidence of sickness. Furthermore — and this is disquieting — a number of industrial concerns are reported to have been compelled to close down at a time when expansion in output is absolutely essential. If the adverse weather conditions had continued for a few more weeks, a large part of the population would have been forced to retire to bed.[5]

The shortage of coal in some areas, however, seems to have been largely a result of distribution rather than production problems. Output in February, for example, at an average of 4,415,000 tons a week, was

40

well above the pre-war figure.

The situation changed again in the spring of 1940. Exports in the early part of the war had been relatively slack, partly through the loss of export markets, but largely because of insufficient shipping to transport the coal. In the spring of 1940 the government sponsored a major coal export drive. The main target was France. The French had recently lost their Eastern coal-fields in the fighting and the British Government determined to make up for the deficiency. A target of 20 million tons of coal a year for France was set. The industry responded by increasing production significantly. In April, May and June production averaged about 4¾ million tons a week, over half a million tons more than the March figure.

The collapse of France in June, however, dramatically changed the position once again. At that time exports to France accounted for about 70 to 80 per cent of the whole of the coal export trade. For the next few months there was a substantial surplus of coal. Certain districts were particularly badly hit. By the first week in July 43 pits were idle and a further 9 partly idle in South Wales. In Durham the position was even worse. The Durham field produced ordinary unscreened coking coal for which it was difficult to find an alternative market. More than 17,000 men left the pits in the area and the Mines Department was forced to introduce measures (a levy on more profitable districts) to support the district finances, for by October a loss of 2s. 2d. was being made for every ton of coal raised.

The downturn in the industry's fortunes created a crisis for the government. It had until that time followed a policy of encouraging men to remain in the mines, yet now it was faced with severe unemployment in the industry. Further, all other industries were working at full capacity for the war effort and the armed services were demanding more manpower. The government reacted to the pressures of this situation and permitted a drift of manpower from the coal-mining industry. The restrictions introduced at the beginning of the war were suspended and by the end of 1940 there were less than 700,000 men working in the industry, compared to 780,000 at the outbreak of war.

This change in the industry's fortunes was once again to prove to be only temporary. By the spring of 1941 there was a shortage of coal and domestic users in the South East were being restricted to 5 cwt. of coal a fortnight. Increased demands for industry and the navy played their part in creating the shortage but the crucial factor was the run-down in the industry's manpower that had taken place over the preceding months. Plans were made to increase production by opening new seams but these had to be shelved because of the shortage of labour. The shortage was most acute in the export districts, South Wales and the North East, and the government was widely condemned for its lack of foresight in allowing miners to leave the industry. The government, in fact, spent the rest of the war trying various methods to increase the level of manpower in the industry. From the middle of 1941 onwards the violent fluctuations in the coal-mining industry's fortunes ceased and the overriding aim of the government was, by some means or other, to increase the output of coal.

Early Measures to Increase Production

The government in mid-1941 took three measures in this direction. The first was the application in May of an Essential Work Order[6] to the industry. The Order acted in three main areas. First, it protected employment in the industry. No man was permitted to leave or to be dismissed from employment in the industry without the prior consent of the government, given through a National Service Officer. Second, every worker in the industry would in future receive a guaranteed weekly wage, whether or not work was available, providing he attended regularly. Third, new disciplinary measures were introduced which might be applied to persistent absentees. Disciplinary action was to be taken by a National Service Officer after consultation with the pit or district production committee.

This Order had a number of important implications and consequences. First, it tied the worker to the industry. In many ways this was nothing new to the mine-worker, for living in an isolated community he rarely had the opportunity to leave even if he had the wish to do so. Yet this Order went further, for it bound him to the industry not through economic or social conditions but through legislation. Second, it led to a recognition of the need to review the wages of mine-workers. The guaranteed weekly wage was clearly a benefit but it was set too low to deal with the real problems. Agreement was reached between the owners and the miners for the introduction of a scheme whereby miners' wages would be increased and they would be offered an incentive for good attendance. The scheme took the form of an attendance bonus of a shilling a shift, payment of which was conditional on attendance at every shift in the week. In fact the scheme did not last very long. It caused so many disputes that the government intervened in September 1941 and suggested that it should be replaced by a flat rate payment. Third, the Order introduced new disciplinary machinery into the industry. Indiscipline was to be dealt with by the state rather than by the employer. A miner in future might not be sacked for indiscipline but he might find himself fined or imprisoned for such an offence. The most important implication of the Order, however, undoubtedly lay in

42

another direction. Its introduction signalled a change in the coal-mining industry's position. The industry had at last emerged from the gloom of the 1930s and the uncertainty of the early war years. Coal was now clearly needed, coal was in demand, and was to remain so for some years to come. Thus Court says:

> The Essential Work Order, as applied to coal mining, marked a new era in mining affairs. The Order was a measure passed to deal with an immediate situation; to ensure that labour was retained in the mines in the summer of 1941 to build up the stocks of coal which would be wanted in the third winter of the war. But it was also the most striking act of recognition by the government and the country of the extraordinary reversal of economic fortune which the war was bringing about in the coal industry. For a generation, coal had been a contracting industry, with too many men trying to live by it and with much unemployment and short time. Compared with the reality of this experience, the activity of the first nine months of the war seemed to many miners a transitory thing; for had it not been followed by a return of the old large-scale unemployment in the export fields after France fell? Incredulous as they might be and were, the miners were from 1941 onwards faced by the fact that a scarcity of labour had set in, which the Essential Work Order helped to prevent growing more acute, but which it could not alter.[7]

The second measure taken by the government in mid-1941 was to introduce a registration scheme for ex-mineworkers. The government and the industry were concerned not only to stop miners leaving, but also to attract ex-miners back to the industry. in June 1941 the Minister of Labour and National Service broadcast an appeal to ex-miners to return to the industry. The disappointing response to the appeal determined the government to take more decisive action. In July of 1941 it introduced the Registration of Miners Scheme.[8] Under the terms of this scheme all men under sixty who had worked for six months or more in the industry since 1935 were to be required to register. On registering they were to be asked whether they would be willing to return to work in the industry (no compulsion was to be used at this stage). A total of 104,000 men registered and nearly a quarter of them indicated that they would be willing to return. The final outcome, however, was not quite so impressive.

One of the reasons for this was the method of administration used. At first the initiative was left with the colliery undertakings to request extra manpower. The poor response from the collieries (in one area it was reported that 10,000 men were available but only 2,000 had been taken on[9]) persuaded the government to change the system. The Minister of Labour, through District Coal Production Committees, directed the returning miners to collieries and it was up to the collieries to use them efficiently. There were, however, numerous practical problems about doing so. Many of the men referred were just not physically suitable for mine work. Thus one mine-owner complained that of 186 names submitted to him by the Ministry only 41 proved to be suitable. Of these only 19 started work and after two days only 16 of them were still working.[10] Apart from physical suitability there were other practical problems such as transport, accommodation, training and the like. The result was that by October 1941 the net increase in manpower was only about 16,000.

The third measure that the government took in mid-1941 was to revive the Coal Production Council. The Council, a body composed of owners, miners and government representatives, had been introduced in April 1940 in an attempt to increase production during the export campaign but had been inactive in the latter part of 1940 and early 1941. In mid-1941 the government decided that the time had come to use it again. The Council was given the task of finding ways of improving production in the mines. To help in this task it had a series of subsidiary district and pit production committees. The Council had a limited degree of success. It succeeded in getting the number of men at the coal face increased and this in turn helped to significantly increase production. The Council has been accused, however, of 'hindering clear thinking'[11] by putting forward a target for the industry in terms of the number of men employed (700,000) when more effort might have been put into other means of raising production. Its subsidiary bodies, the pit committees, have also been described as 'lamentably unsuccessful'.[12] Two reasons were suggested for this. First, they concentrated too much attention on absenteeism; in many cases they became little more than absenteeism tribunals. Second, in many cases the management refused to accept the combined discussion of production problems. They saw this as interference in managerial prerogatives.

1942 White Paper

Despite these measures the position deteriorated in the winter of 1941-42. By the spring of 1942 there were fears of another 'coal crisis'. Estimates were presented to the Cabinet which suggested that there could be a serious coal shortage by the winter. Even though the estimated coal requirements had been severely pruned they still stood at 215 million tons for the year ending 30 April 1943. The likely maximum production of coal, during the same period, however, was just over 205 million tons. In June 1942, therefore, the government published a White Paper[13] outlining their plans for dealing with the coal crisis.

The White Paper opened with a statement of the basic problem:

> Owing to the expansion of war production and other war-time causes the demand for coal is still increasing. But the output of coal is tending to decline. The present rate of production is not yielding enough coal to cover unrestricted domestic and industrial demands. It is therefore necessary that immediate steps should be taken to increase the production and to eliminate unnecessary consumption of coal.[14]

The remedies can be subsumed under three headings; manpower, organisation and consumption. The government proposed, in an attempt to improve the manpower position, to bring a number of ex-miners back from the armed forces. It suggested that a total of 6,500 from the Army and 1,300 from the RAF would be involved. The White Paper recognised, however, that such a move was a once and for all affair and could not be repeated year after year. To deal with the long-term problem

> More lasting remedies must be found for closing the gap between normal entry into, and exit from, the industry by increasing the entry of boys and youths, and checking the outflow of older men.[15]

The mines were to be added to the list of priority industries into which men might go instead of joining the armed forces. The problem of juvenile recruitment was to be dealt with by a committee under the chairmanship of Sir John Foster. (The report of the committee published later in 1942[16] recommended improved training, welfare facilities and promotion prospects as means of increasing juvenile recruitment.)

The manpower position, it was suggested, might also be eased by a reduction in the level of absenteeism. Absenteeism had long been a contentious issue in the industry. The coal-owners had frequently criticised the miners for their 'high rate' of absenteeism, arguing that absenteeism had increased during the war because of the increase in the level of earnings. Wilson, however, has suggested that the coal-owners' argument obscured the real position.[17] There had, he agreed, been an increase in absenteeism during the Second World War; absenteeism, for example, was 10.4 per cent in 1942 compared to

6.9 per cent in 1939. At the same time, though, there had been an increase in the number of shifts worked per man per week, from 5.15 shifts in 1939 to 5.34 shifts in 1942. The apparent paradox is explained by the way absenteeism was calculated. It was calculated on the basis of the number of shifts lost as a percentage of all possible shifts in that week. Between 1939 and 1942 the number of shifts on which work was available had increased, so had the number of shifts worked, but not in proportion. Thus between 1939 and 1942 the rate of absenteeism and the length of the working week had both increased.

This line of argument was broadly accepted by the White Paper. It said:

> As regards absenteeism, there has been some tendency in recent discussions to exaggerate the gain in production which might be secured in checking absenteeism and by additional effort on the part of the men. It is doubtful whether in fact avoidable absenteeism is greater in the mines than in other industries. While the absenteeism percentage has risen substantially during the war, there has also been a steady rise in the number of shifts worked per wage earner; and in the last week for which statistics are available the number of shifts worked was 5.66, the highest on record. Charges of excessive absenteeism cannot be sustained against the great majority of miners.[18]

Nevertheless the government had decided that further measures should be taken to eliminate what avoidable absenteeism existed. To this end it proposed to strengthen existing provisions for dealing with the problem. Previously most cases of absenteeism had been dealt with by the pit committees who had the power to refer a man to the National Service Officer for prosecution, but rarely used it. The government announced in the White Paper that it was going to appoint Regional Investigation Officers to whom all cases of absenteeism, persistent lateness and other disciplinary offences were to be referred.

The government's measures announced in the White Paper to reorganise the industry had a great deal in common with those introduced during the First World War. They were based on the notion of 'dual control'. This meant that the government assumed supervision of the industry's affairs but left the ownership in private hands. More specifically, the mines were placed under the supervision of a government Minister who was assisted by a Controller-General and a National Coal Board. This latter body was under the chairmanship of the Minister and had amongst its members the Controller-General, Regional Representatives of the industry, a number of pit managers and colliery technicians, and representatives of coal distributers and consumers. Much of the day-to-day work of the industry, however, was delegated to regional machinery. This was under the direction of a Regional Controller assisted by a Regional Coal Board (constituted on a similar basis to its national counterpart). Pit production committees continued in operation helping to secure maximum output but they were relieved of their responsibilities for absenteeism.

One important item over which this new machinery was not to have jurisdiction was wages. The government took the view that 'the success of the proposed National Coal Board as a body for promoting increased production would be gravely prejudiced if it were associated in any way with wages questions.'[19] It did, however, recognise the need for some national machinery to deal with wages matters. Thus:

It is, however, desirable that a system should be developed by which questions of wages and conditions in the mining industry would be dealt with on a national basis and by a properly constituted national body.[20]

The third area dealt with by the White Paper was the consumption of coal. The government outlined in the Paper a scheme for rationing domestic coal, which had been devised by Beveridge and was based on points awarded for various types of 'need'. The White Paper stated, however, that 'the Government have decided that it is not essential that this scheme should be introduced forthwith.'[21] In fact, the proposed scheme was not introduced during the period of the war at all.

The White Paper was given a mixed reception. Its proposals to increase manpower were broadly welcomed, but its scheme for rationing was attacked by many Conservative MPs. It was this opposition which was largely responsible for the scheme being abandoned. Most attention, though, was necessarily focused on the proposals for government control of the industry. Neither the Labour Party and the miners on the one hand, nor the coal-owners on the other were entirely satisfied. The Labour Party and the miners had argued for the nationalisation of the industry; the coal-owners disliked any kind of government intervention in the industry at all and argued that such intervention did little to raise output during the First World War. Yet, at the same time, all parties recognised that the government's plans gave them at least a little of what they wanted. The industry had not been nationalised as the Labour Party and the miners had hoped but at least the government had taken some steps to control the industry. The coal-owners had not been spared government intervention altogether but at least the intervention took the form of 'dual control' rather than nationalisation. The result was that the government's plans were grudgingly accepted by both sides. In the House of Commons the White Paper was approved by 329 votes to 8.

Negotiating Machinery

The White Paper had deliberately avoided involving the new National Board in any controversy over wages. This did not, however, avoid the wages issue being raised in mid-1942. The Miners' Federation argued that wage increases were essential not only because the standard of living of their members had fallen but also to ensure the success of the government's policy of attracting labour to the mines. Accordingly they asked for a general wage increase of 4s. 0d. per shift and a national minimum weekly wage of £4 5s. 0d. The owners were not unwilling to concede a wage increase but they objected to the basis of the miners' demands. Their objections concentrated on two points; first, they argued that any wage increase should be linked to increased production; second, they stated their total opposition to a national minimum wage. In an attempt to avoid a stoppage of work the government appointed a committee to inquire into the dispute.

The Board of Inquiry was appointed with Lord Green, the Master of the Rolls, as chairman. Its terms of reference were

1. To consider and report in the first instance upon the immediate wages issue; and further
2. To inquire into the present machinery and methods of determining wages and conditions of employment in the industry, and to submit recommendations for the establishment of a procedure and permanent machinery for dealing with questions of wages and conditions of employment in the industry.[22]

The Board produced four reports in all.

The first report[23] was published later in June 1942 and dealt with the immediate wages issue and two related topics. Its recommendations went some, but not all the way, towards meeting the miners' claims. It argued that a wage increase of 4s. a shift, unrelated to productivity, would be excessive and suggested instead increases limited to 2s. 6d. a shift; it accepted the need for a national minimum wage, although the figure suggested was £4 3s. instead of the £4 5s. asked for.

Two other matters were dealt with by the report which had not formed the basis of the miners' claim. First, the report argued that its award should lead to a 'stable' increase for miners. It was possible under existing arrangements for the award made to be reduced because

48

of the practice of varying national wages to take account of district conditions. The report suggested that this procedure should be amended so that wages could not, in any district, go below the level they had recommended. Second, the report suggested a productivity bonus. The bonus was to be over and above the general award and was not an acceptance of the owners' original proposals.

The government accepted the Board's recommendations on the wage increases and accepted their recommendations on bonus payments in principle. However, they asked the Board to consider the question of bonus payments further and present more detailed recommendations.

Although the report did not meet all of the miners' original claims it went a good way towards doing so and seemed to persuade the miners that at least they were getting more sympathetic attention. They were particularly pleased with the Board's proposals on the minimum wage. The President of the Miners' Federation, W. Lawther, said that as a result of the Board's recommendations, 'all legitimate grievances in relation to the minimum wage advance had been removed.'[24] The immediate threat of industrial action thus receded and the Board turned to consider other issues.

The Board presented its second report[25] in August 1942. The report dealt in detail with the issue of bonus payments. The Board declared its own preference for a scheme based on pit productivity. The two sides of the industry differed and both suggested a district assessment. In the end the Board agreed to accept the wishes of the two sides because it felt that their good will was essential if the scheme was to work. It did so, though, on the understanding that if the scheme failed to live up to expectations then the notion of pit assessments would be revived. The scheme provided for maximum bonus payments of 3s. 9d. a shift on attainment of output in excess of the target (based on previous production) by 15 per cent.

The Board's third report[26] was not presented until March 1943. This dealt, however, with the more crucial issue of conciliation machinery for the industry. The report argued that the special characteristics of the mining industry meant that it would not be possible to 'transplant' machinery from another industry. They therefore set themselves the task of devising completely new machinery, which took the form of the National Conciliation Board.

The National Conciliation Board itself was to consist of two bodies, the Joint National Negotiating Committee and the National Reference Tribunal. The Negotiating Committee was to be composed of twenty-two members, half nominated by the Mining Association of Great Britain and half nominated by the Miners' Federation of Great Britain. All questions which came within the scope of the scheme would normally first be discussed by the Negotiating Committee and this committee alone would try to resolve the matter. If, however, an agreement was not reached within five weeks then the matter would be passed on to the second body, the National Reference Tribunal. The Tribunal was to be composed of three permanent members, none of whom were to be either engaged in the coal-mining industry or Members of the Houses of Parliament (except as a consequence of holding judicial office). The members were to be appointed by the Master of the Rolls after consultation with the Mining Association and the Miners' Federation, and serve for a period of five years. When the tribunal heard cases it could have the help of assessors suggested by the two sides of the industry. The Tribunal would only normally consider matters referred to it by the Negotiating Committee but there would be certain matters which would be within its exclusive jurisdiction (such as those concerning the conciliation scheme itself or an award the tribunal had previously made) and other matters could be referred direct by the Minister of Fuel and Power (although the Tribunal would be required to consider the views of the Negotiating Committee on such matters). The decision of the National Tribunal was to be final and binding on all parties.

The report also suggested that district conciliation machinery should be established to deal with purely local questions. An opportunity should first be given for the two sides of the industry to set up this machinery themselves, but if they failed to do so then the National Tribunal should assume the responsibility. In such a case the machinery should be closely modelled on its national counterpart and consist of a District Negotiating body with a District Referee to arbitrate should agreement not be reached in the negotiating body. The report also suggested that provision should be made for the reference of important matters from the district to the national machinery for consideration.

The Board's third report was in many ways its most important; it certainly had the greatest long-term impact. The conciliation machinery which was established for the industry was based on the report's recommendations and it was machinery which was to last far beyond the wartime period. It lasted, in fact, without any major alteration for a period of about twenty years.

The kind of machinery suggested undoubtedly had many virtues but to the mineworkers one was paramount. They had at last achieved comprehensive conciliation machinery for the industry on a national basis. Previously the owners had always refused to consider national machinery arguing that disputes were better settled on a local basis. They had gone some of the way towards accepting the principle of national machinery after the wages dispute of 1935, but that machinery had never been brought into operation satisfactorily. The suggestions of the Greene Board went far beyond what the miners had achieved before and they were clearly delighted.

The Board's fourth and final report[27] was published in September 1943. This report reviewed the operation of the bonus scheme which had been introduced in September 1942 following the Board's earlier report. It suggested that the scheme had worked satisfactorily for the first few months but had not been doing so since. There were, it suggested, a number of possible reasons for this, such as the impact of local strikes and absenteeism. Consequently the Board suggested a revision of the scheme. Two options were put forward for consideration, a scheme based purely on pit output and a scheme based on both district and pit output. The Board indicated that it felt the former scheme would be the most useful although either would be an improvement on the existing position.

Discussions immediately commenced on the Board's proposals. The government took the view that the pit scheme offered the greatest advantages and proposed such a scheme in November 1943. In fact, the scheme was never introduced. The two sides of the industry opposed its introduction and it was eventually discarded.

Further Problems

By the autumn of 1943 the government was faced with yet another 'coal crisis'. Despite all the measures it had taken, the return of ex-miners, the introduction of 'dual control', stricter penalties for absenteeism and the output bonus, the problems of achieving a satisfactory supply of coal had still not been solved. The problems arose from a number of different sources. First, there were renewed fears about the manpower situation. The recall of ex-miners from the forces had more or less been completed yet wastage was still outstripping normal recruitment. Second, output per man engaged was falling. In the third quarter of 1943 output per man employed was nearly 9 per cent less than the comparable figure for 1939. Third, developments in the war meant that more coal was needed for the war effort. In summary, then, the problem in late 1943 was one of falling output but increasing demand.

Speaking at the Miners' Federation Conference in 1943, Bevin, Minister of Labour, told delegates how serious the position in the industry was.

> At the end of the year there will not be enough men and boys in the industry to carry it on. It is a serious position. It is one of the great difficulties of the war effort.[28]

He concluded by saying,

> It is obvious that I shall have to resort to desperate remedies during the coming year. I shall have to direct men to the coal mining industry.[29]

In December 1943, therefore, the government introduced conscription for the mines.[30] As a result, when men became available for service with the armed forces they were likely to find themselves required to work in the mines instead. Conscription to work in the mines was to be operated on a ballot basis so as to ensure random selection. It was intended that the ballot should provide 50,000 new recruits for the industry in 1944.

In fact, the scheme was not nearly as successful as had been hoped. The number of 'Bevin boys' (as they became known, after the then Minister of Labour, Bevin) who entered the mines was only 15,000 in 1944 and between 5,000 and 6,000 the following year. Thus, in 1944 'Bevin boys' formed little over 25 per cent of entrants to the industry and in the following year only about 15 per cent. There were a number of reasons for this apparent failure, but two were particularly important. First, there were a number of practical problems, just as there had been earlier with the volunteers, such as accommodation and transport. Second, there was the problem of training. An official scheme of six weeks training was devised but the success of the training depended almost entirely on the local pit officials and managers.

Renewed Wages Problems

Towards the end of 1943 the miners' leaders began formulating new wage demands. Miners had improved their position in the wages 'league table' with the 'Greene Award' of 1942, but by the end of 1943, much of this advantage had been lost. Immediately after the Greene award the miners had occupied twenty-first position in the earnings league table, but eighteen months later they had fallen to forty-first position. Stressing the need to attract men to work in the mines the Miners' Federation put in a wage claim for new national minimum wages of £6 for underground workers and £5 10s. for surface workers. In addition, they asked for a national agreement covering overtime rates and increases in piece rates. The claim was eventually submitted to the new National Reference Tribunal, whose chairman was Lord Porter.

The Tribunal published its report in January 1944.[31] It accepted the need to increase minimum rates in an attempt to make the industry 'more popular' and suggested new minima of £5 a week for underground workers and £4 10s. a week for surface workers. It also agreed to the introduction of new overtime rates. However, it refused the claim for an increase in piece rates. In conclusion the Tribunal said it looked on the award as

> merely a temporary expedient which will afford an opportunity for the wages structure throughout the industry to be reconsidered and thoroughly reviewed in conjunction with the general conditions obtaining in it.[32]

The miners' reaction was mixed. In a number of districts there was disappointment at the award which found expression in strike action. In the first week of February 1944 178,700 tons of coal were lost as a result of stoppages of work, the highest weekly figure since the beginning of the war. Reaction at the national level, though, was not so hostile and the Miners' Federation accepted the award with the proviso that negotiations should begin to remove some of the anomalies it had created.

The discussions on the anomalies began in the Joint National Negotiating Committee and it was quickly agreed that although some action would have to be taken it would be better taken on a district than a national basis. Accordingly negotiations were started in individual districts.

The position of the government in relation to the industry's finances at this stage was crucial. The government had played an indirect role in the finances of the industry ever since the start of the war. An agreement had been reached between the government and the owners at the beginning of the war that coal prices would not be raised without government approval and a scheme had been introduced in 1940 to assist coal-fields which had been damaged by the loss of important export markets. When the guaranteed wage provisions were introduced along with the Essential Work Order in 1941 the government intervened once more to impose a levy on the industry to meet the cost of the guarantee.

In 1942 when the government took over control of the industry these arrangements were suspended. The government did not intervene directly in the industry's finances but instead brought in the Coal Charges Account. Under this scheme a new levy was introduced, initially of 7d. per ton of coal produced. The money received was administered jointly by the Minister of Fuel and Power and the Treasury and could be used 'for any purpose connected with the production or marketing of coal.'[33] In fact it was primarily used to subsidise the increased cost of wartime production. Despite the frequent increases in the levy (by 1945 it had reached 15s. a ton) and a government declaration 'that all costs of the industry should be met in full by the Industry without Exchequer subsidy,'[34] by the end of the war the account was in debt to the tune of £25 million, a debt which was paid by the Exchequer.

The government after 1942, then, had effective control of the organisation of the industry and the price of coal. Further, through the Coal Charges Account it subsidised the costs of production. The result was that the government in fact, if not in theory, exercised great influence in wage matters. If, for example, the miners were given a wage increase by the mine-owners, then the government had to decide how this should be paid for. Should it be paid for by increasing prices, or should it be paid for by increasing the government subsidy to the industry? If it was not to be paid for in either of these ways, then could the wage increase really be given? It was this kind of question which arose in 1944. The government eventually agreed to cover the original Porter award through the Coal Charges Account and there was apparently a belief that the government would also cover any further increases negotiated in the districts to remove the anomalies. Consequently, employers and unions quickly agreed on substantial wage rises and increases of about 15 per cent on piece rates, for example, were agreed in South Wales. In fact, the government was by no means committed to covering any such wage agreements. They claimed that their intentions had either been misrepresented or misunderstood. The Minister of Power gave his recollection of what he had told the leaders of the Mining Association about the wage rises in the House of Commons on 15 February 1944.

> On February 3rd, 1944, without prior consultation with my Department, the Joint National Negotiating Committee agreed that the anomalies arising out of the Porter Award should be adjusted by the two sides in each district, provided the cost of such adjustment was met from the Coal Charges Account. Later the same day representatives of the Mining Association informed my officials of this agreement and were at once told that no undertaking could be given that the cost of such adjustment would be paid from the Coal Charges Account and that important issues were being raised which would require consultation between me and my colleagues. On the following day I myself saw representatives of the Mining Association and confirmed this statement of my officials, and also intimated that I was not prepared, without consulting my colleagues, even to agree that any sum paid in order to implement the Porter Award itself, quite apart from the adjustments of the anomalies, would be refunded from the Coal Charges Account. The

Mineworkers Federation on February 4th, 1944, was informed
verbally of the answer given by my official on the previous day to
the Mining Association, and I understand that, before any district
negotiations took place, the Mineworkers' Federation made it
clear to its district officials that any agreement that might be
reached as to the adjustments of anomalies would be subject to
His Majesty's Government agreeing that the cost thereof should be
met from the Coal Charges Account.[35]

Whatever the cause of the misunderstanding the result was clear.
Mine-owners could no longer afford the agreements they had discussed
with the unions and therefore withdrew their offers. The reaction was a
series of damaging strikes throughout the industry. In the first quarter
of 1944 over 2 million tons of coal were lost through labour disputes
(see Table 7), double the amount lost in the whole of the previous
year. Again, during the first three months of the year 1,641,000 man
days were lost as a result of strikes in the industry, about 75 per cent
of the man days lost through strikes in the country as a whole.
    Early in March the government called the two sides together to
discuss the situation and stressed the extreme urgency of the position.
A considerable amount of coal was being lost through stoppages, enough
in fact to wipe out the effect of all the government's measures to
increase coal production. The result of the meeting was an agreement
for a restructuring of wages in the industry. The final agreement was
complicated but essentially it attempted to do two things. First, it
tried to amalgamate as far as possible all of the various supplements
and additions to wages into the basic wage. Second, it aimed to
increase piece rates so as to give miners an incentive to produce more.
    The remainder of the war was far from tranquil as far as the coal
industry was concerned. The winter of 1944-45 saw severe coal
shortages. These were partly the result of increased demand abroad,
partly the result of the winter weather, and partly the result of
inadequate production. In addition the supply of coal was severely
hampered by serious distribution difficulties. The attention of all
sides, however, had in many ways turned away from immediate
wartime problems and debate ensued on the future of the industry
after the war.

Towards Nationalisation

Haynes, in his review of the nationalisation of the coal-mining
industry,[36] emphasised that the industry prior to nationalisation
faced not one but a whole variety of problems. He selected nine of
these problems for particular emphasis. The first was the depletion of

the coal reserves. Britain's coal seams were not in 1945 in danger of being worked out entirely; rather there was a gradual deterioration in their quality. The cost of producing coal in Britain was rising because of the need to work thinner seams at greater depths and under more difficult conditions. Second, low productivity was a problem. In 1938 output per man shift in Britain at 1.148 statute tons was considerably lower than that in the USA (4.73), Poland (1.787), the Netherlands (1.619), Czechoslovakia (1.424 in 1937), the Ruhr (1.523) and Upper Silesia (1.830). Only the Saar (1.121), France (0.820) and Belgium (0.741) had a lower output per man shift than Britain. It was suggested that in terms of natural conditions coalfields in Britain might best be compared with those in Holland and the Ruhr. Third, and one of the main reasons for the low output per man shift, was the problem of mechanisation and mine layout. Haynes quoted evidence from the Reid Report (which will be referred to later) to support the contention that qualitatively if not quantitatively the extent of mechanisation in British mines was unsatisfactory. It was also suggested that the British industry had concentrated too heavily on certain types of mining, in particular the Longwall Advancing system of mining, to the exclusion of the Room and Pillar system and the Longwall Retreating system. Fourth, there was the lack of integration in the industry. Attempts had been made by the government in the 1920s and 1930s to encourage amalgamations but with little success. The result was that in 1945 there were still too many small concerns operating in the industry and there was too little 'over all' organisation. Fifth, there was the declining quality of coal. This was largely the result of previously mentioned factors; the poor quality of the seams, poor methods of obtaining coal, lack of integration in the industry, and so on. The result was that domestic and industrial consumers in Britain were not receiving the type of coal they were used to, while the deteriorating quality of the coal undoubtedly damaged Britain's export trade. Sixth, there were the problems caused by the instability in the sale and price of coal in the inter-war years. One of the most serious consequences was in the area of employment. In the inter-war years the industry had been plagued with unemployment (for the greater part of the period above 15 per cent of insured workers were unemployed) and this in turn caused much of the distrust and bitterness then evident amongst coal-miners. Seventh, there were the problems of the underproduction of coal and the underrecruitment of workers in the industry. The problem of underproduction, he suggested, was itself linked to a number of other issues, like the rise in the average age of the miner, increased absenteeism and so on. The problem of underrecruitment was one which had come to the fore during the war but which had been present for some time before. Young people were not entering the industry, partly because of the poor conditions and partly because of increased opportunities elsewhere. The Foster Report[37] had shown that in 1942 only 14,000 juveniles had entered the industry compared to 30,000 in 1934. Eighth, there were the problems concerned with the working and living conditions of the coal-miner. The coal-mining industry, for example, had one of the highest accident rates; in 1945, 550 workers were killed (506 underground) and over 2,300 seriously injured. To this has to be added the deplorable housing conditions and amenities in many pit villages. Finally, there was the question of labour relations. There was, Haynes said,

little question that labour relations were less satisfactory in the coal trade than in any other major industry in Great Britain.[38]

The Reid Report

Many of these problems were highlighted in an official report[39] published in March 1945. The report of the committee established under the chairmanship of C. Reid in effect confirmed many of the worst fears of contemporary commentators about the coal-mining industry and as a result was to have a crucial role in the nationalisation debate.

The committee's terms of reference were:

To examine the present technique of coal production from coal face to wagon and to advise what technical changes are necessary to bring the industry to a state of full technical efficiency.[40]

After a review of the history of coal-mining in Britain, the committee in its report turned to a comparison of the position of the industry in Britain with that in other countries. It pointed out that over the previous ten years or so output per man shift had increased in most other countries far faster than it had in Britain. Taking countries that had broadly comparable natural reserves, output had increased by 54 per cent in Poland (1927-36), 118 per cent in Holland (1925-36), 81 per cent in the Ruhr (1925-36) and by only 14 per cent in Britain (1927-36).

Some of the other countries, of course, had a great deal of headway to make up because of their relatively late entry into the field. This could not, however, account for all the differences noted and it was clear that the British industry had not taken sufficient advantage of new technical developments. A whole range of issues were mentioned but particular emphasis was laid on improvements that might be made in haulage. No single operation, the report argued, offered more scope for improvements in efficiency; the application of new locomotive haulage was essential. In Britain, for example, on average 5 tons of coal was handled per haulage worker, whereas in Holland the total was between 20 and 25 tons and in USA 50 tons.

Why, then, had Britain failed to take advantage of these new developments? Again, many reasons were cited but two were afforded prominence. First, there had been inadequate investment in the industry. This had been because

> The British industry as a whole has been in a perpetual state of financial embarassment, and the long-standing uncertainty surrounding the future ownership of the industry has not been conducive to expenditure on long-term improvements necessary to raise the general efficiency of the industry.[41]

Second, there was the question of the organisation of the industry. In Britain there had been no real attempt to reorganise and rationalise the industry to allow for more efficient production comparable to that in other countries.

> The grouping of a number of mines under the same ownership on the continent has facilitated the closing down, or merging, of uneconomic mines, and the concentration of operations to the remaining shafts. In Britain, ownership is widely dispersed, and this avenue to greater efficiency has never been explored on any adequate scale.[42]

This second reason for Britain's failure to take advantage of the new developments was given prominence in the report's conclusions. The committee clearly linked the future of the industry to technical improvements, and these in turn to reorganisation.

> We have now come nearly to the end of the task which we undertook. We have analysed the general causes of low productivity of British mines, showing the difficulties under which the mining engineer has laboured, and we have made drastic and far-reaching recommendations for technical changes. We have carefully considered whether we have thereby fulfilled our duty. As a result we have come to the conclusion that it is not enough simply to recommend technical changes which we believe to be fully practicable, when it is evident to us, as mining engineers, that they cannot be satisfactorily carried through by the industry organised as it is today.[43]

58

This industry needed to be reorganised, the committee argued, by some strong central authority.

> If a comprehensive scheme of reorganisation along the lines we have recommended is to be carried through, we consider that an Authority must be established which would have the duty of ensuring that the Industry is merged into units of such sizes as would provide the maximum advantages of planned production, of stimulating the preparation and execution of the broad plans of reorganisation made by these units, and of conserving the coal resources of the country. The existence of such an Authority, endowed by Parliament with really effective powers for these purposes, is, we are satisfied, a cardinal necessity.[44]

The Reid Committee Report had a substantial impact. It had clearly analysed the technical deficiencies of the British coal-mining industry. More than that, however, it had suggested that the technical improvements that were essential could not be brought about unless there was a complete overhaul of the industry's structure. The report stopped short of recommending nationalisation but its conclusions proved to be a potent weapon in the hands of those who supported such action.

The Future of the Industry: Different Views

By the end of the war there were a number of different views about future policy for the British coal-mining industry. All agreed with the Reid Report that changes needed to be made in the structure of the industry; they differed on how the changes should be brought about and what kind of structure of ownership and control should emerge.

The clearest statement of the coal-owners' views is given in the 'Foot Plan'.[45] This was a plan for the industry in the post-war period drawn up by Mr. Robert Foot (formerly Director-General of the BBC) 'independent'[46] chairman of the Mining Association. The plan was based on twenty principles:

1. There should be a national conception of the industry's position and responsibilities.
2. Mining should be undertaken in accordance with the best known principles and practice.
3. The best possible professional advice should be made available to all undertakings.
4. The industry should be brought to and maintained at the

highest possible state of efficiency.

5. There should be a closer integration of the various undertakings in the industry.
6. No existing pit should be closed unless the workers displaced could be absorbed elsewhere and no new pit should be opened unless there were satisfactory social arrangements for the workers.
7. A more efficient method of selling and distributing coal should be introduced.
8. The workers in the industry should be guaranteed a good standard of living and stability of employment.
9. There should be an adequate return on capital employed.
10. There should be full information for the workers and close liaison between the two sides of the industry.
11. Existing conciliation machinery should be maintained.
12. The best principles of labour administration and management should be applied throughout the industry.
13. Every colliery should apply the best safety practice.
14. Every encouragement should be given to research.
15. Proper arrangements should be established for the recruitment, education and training of new entrants.
16. The workers of the industry should have the right to call these principles to their aid in any matter in which they could show that they were not being kept fully in operation in any undertaking.
17. Contact should be maintained on all matters affecting national policy with the Minister of Fuel and Power.
18. Close contact should be maintained with the Coal Commissioners.
19. Provisions should be made to aid colliery finance.
20. Subject to the above principles there should be no interference with the autonomy of any undertaking.

The Foot plan also proposed that

> for the proper observance and maintenance by the industry of these principles (and of any others that might be established from time to time as governing principles for the operation of the industry) and for the purpose of acting as Trustees to Parliament and the country at large for such observance and maintenance[47]

a central authority should be established called the 'Central Coal Board' consisting of people who could 'be relied upon to approach their responsibilities in a statesmanlike way to think nationally not sectionally'.[48] In fact, the Central Board was to be dominated by coal-owners and their nominees. He also proposed that the Central

Board should be assisted by a series of district coal boards who would act as 'outposts' of the Central Board and assist in maintaining the principles.

The Foot Plan, in recognising the need for the industry to be fundamentally restructured, went further than the owners had done before. Nevertheless, the plan retained two ideas which had dominated the actions and the attitudes of the coal-owners throughout the twentieth century. The first was that of voluntary action. There was no suggestion of legally enforceable powers to carry amalgamation proposals through. The plan essentially proposed a meeting of colliery owners, called the Central Board, to work out voluntary amalgamation and reorganisation schemes. The second was that of private enterprise free of state control. The industry would liaise with the government but the government would have no special jurisdiction over it.

The views of the Conservative Party are not quite so clear. There is, in fact, often an element of confusion about their point of view because of a speech made by Churchill in the House of Commons in 1943. In this speech Churchill appeared to accept the wisdom of nationalisation when he said,

> Take the question of nationalising the coal mines. These words do not terrify me at all . . . the principle of nationalisation is accepted by all providing proper compensation is paid.[49]

These words, however, should not be taken out of context for it is clear when one reads the whole of his speech that what Churchill was accepting was the principles of nationalisation for some industries in Britain in certain circumstances. He was not accepting the need or the desirability of nationalisation for any particular industry, and certainly not for the mines.

Undoubtedly a clearer statement of Conservative Party thinking can be seen in a document produced by the Tory Reform Committee.[50] This document envisaged the appointment of a Reorganisation Commission consisting of three Coal Commissioners (independent persons), three representatives of the Mining Association and three representatives of the Miners' Federation, to oversee the reorganisation of the industry. Mine-owners would be required to submit schemes for voluntary amalgamations to the Commission within a certain period of time. If satisfactory schemes were not produced on a voluntary basis then the Commission would be empowered to produce schemes itself. It was envisaged that such amalgamations would initially reduce the number of undertakings operating in the industry to about two hundred and eventually to sixty or forty. The plan also proposed the establishment of what it termed a 'Miners' Charter'. The Charter would provide for security of employment, minimum earnings and the like.

In so far as this plan envisaged compulsory amalgamations it went beyond that of the coalowners. Nevertheless, the authors were careful to state their belief in the virtues of private enterprise and argued, in fact, for greater rather than less freedom, to be given to the new larger coal concerns. Thus, the plan said:

> The Government's stimulation of coal reorganisation and development should therefore be accompanied by a relaxation of Government interference in every aspect of day-to-day management. The sole responsibility for management should be restored unequivocally to colliery managers.[51]

By the end of the war the Conservative Party and the coal-owners' views on the future of the industry had moved very close indeed. The coal-owners effectively abandoned their plan and accepted that put forward by the Tory Reform Committee. The Conservative Party, for their part, in the election manifesto of 1945, declared their belief in the value of free enterprise for the coal-mining industry. Their aim, they said, was 'to preserve the incentives of free enterprise and safeguard the industry from the dead hand of state ownership or political interference in the day-to-day management of the mines.'[52]

Socialist opinion in general had for a long time supported the idea of the nationalisation of the mines. The Labour Party, the TUC and the miners had all declared their support for such a course of action by the turn of the century. Socialist opinion was not so consistent, however, on the form nationalisation should take. Initially it seemed to favour at least an element of workers' control (the extent of workers' control favoured is discussed in an earlier chapter). This view found expression in the Bill presented to Parliament on the miners' behalf in 1919; it was based on the scheme outlined to the Sankey Commission and was reintroduced in 1923, 1924 and 1925. The view also found expression in the Labour Party's proposals in their programmes 'Labour and the New Social Order' and 'Labour and the Nation'.

By the mid-1920s, however, there was evidence that this point of view was losing ground. In their evidence to the Samuel Commission both the TUC and the Labour Party modified their earlier proposals for the industry. Control was to be diffused over a number of separate bodies, only some of which would contain representatives from the miners and their unions, and considerable effective power was to be given to the Secretary for Mines. Further moves in this direction were indicated in a memorandum on the administration of the industry under nationalisation prepared in 1928 by Shinwell and Stacey. In this memorandum the authors suggested that

> a comparatively new and valuable approach to the problem may be found in regarding the question purely and simply as a business one . . . After all, the first essential for any scheme of nationalisation is that it should work efficiently.[53]

The Miners' Federation would not be directly involved in operating the new machinery; the only element of 'workers' control' would 'arise from the fact that all the officials . . . would naturally and inevitably be recruited from the industry itself.'[54]

Although the memorandum was rejected by the miners, as Hanson[55] says, a process of conversion was taking place in the late 1920s and in succeeding years. The miners, the unions and the Labour Party all tacitly or explicitly dropped the idea of 'workers' control' of the industry after nationalisation. In 1931, for example, the Labour Party accepted Morrison's 'efficiency concept of nationalisation' which argued that nationalised industries should be run by the 'best brains that we can secure'. Again, the TUC in their plans for the post-war reconstruction of industry argued for the maintenance of trade union independence. Trade unions might nominate people to serve in the administration of a nationalised industry but anyone appointed would not be a trade union representative and would be required to relinquish his trade union membership on appointment. Perhaps the best example, though, of the way a great deal of socialist opinion had moved away from the notion of "workers' control" by the end of the Second World War can be seen by reference to one of the leaders of the Guild Socialist movement in the 1920s, G. D. H. Cole. (He had, in fact, played a large part in preparing the miners' scheme of 1919.) Commenting specifically on the mining industry, Cole, at the end of the war, argued against any form of "workers' control" at that time. Eventually, he suggested, workers might be able to take a fuller part in the government of the industry, but at the end of the Second World War such a method of control would be inappropriate.

> Why, it will be asked, if the Guild Socialist solution is best in the long run, will it not do now? The answer is easy. The Guild method implies the existence among all those who work in the industry, from administrators and managers to unskilled workers at the pithead, or at any rate amongst those of them who have most influence among their fellows, of an attitude of responsible acceptance of the obligation to put the public interest in the first

place. Such an attitude can be expected in any high degree only in a society permeated through all its parts by the spirit of justice and equality, set free from class conflict, and conscious that rights and duties go together and that the obligation of service is the correlative of freedom. Moreover, such an attitude cannot in the nature of things come existence suddenly. Men are creatures of habit; and neither among managers nor among workers can the habits of mind established under capitalism be suddenly transcended as soon as an industry passes from private to public ownership. The industrial relations character of capitalist society are bound to leave an evil legacy behind; and in the coal industry, with its exceptionally bad record of relations between masters and men, this legacy is bound to be worse than in most others.[56]

By 1945, then, there was a strong movement for a radical reform of the industry's structure. The return of a Labour Government made it inevitable that the radical reform would take the form of nationalisation. All of the parties involved may not have welcomed the move — some, like the coal-owners, violently opposed it — but most were resigned to it.

The formal announcement of the new government's intention to nationalise the mines was made in the King's Speech in April 1945. A Bill was introduced into Parliament by the end of the year and it received the Royal Assent in July 1946. Vesting day for the industry was fixed at 1 January 1947. After they had abandoned the idea of 'workers' control' the Labour Party had for a while favoured the idea of nationalised industries being run as part of the government machine, but such ideas had long since been abandoned; the industry was to be run by a public corporation, the National Coal Board, independent from the Civil Service but responsible to Parliament through the Minister of Fuel and Power.

The National Union of Mineworkers

It is appropriate at this juncture to look in more detail at the mine-workers' unions. Throughout the twentieth century, up to the Second World War, most mine-workers were represented by a body known as the Miners' Federation of Great Britain. This body formed in 1888 was as its title suggests a Federation of unions, not a union in itself. Miners were, in fact, organised into a vast number of small local trade unions. Many of the unions had joined together in district associations, but even these district associations were fairly loose affairs. The Federation was run by a National Executive Committee (elected each year at an annual conference), and a President and

64

General Secretary, but neither permanent officers nor the national executive could overrule an individual union. In 1942 the Federation was described by its President as a 'bow and arrow type of organisation'.[57]

Some moves were made during the 1930s towards strengthening the district and national organisations of the unions. Attempts made to improve district organisation, for example, might be noted in South Wales in the early 1930s, and in 1937 the annual conference of the Federation accepted a resolution which put forward the principle of one mine-workers' union. Despite such moves the Miners' Federation remained, until the end of the Second World War, true to its name, and merely a coordinating body.

It is impossible to offer a complete explanation for such a method of operation, yet one important factor might be mentioned. There is little doubt that the miners' unions were strongly influenced by the structure of collective bargaining in the industry. Throughout most of the 1920s and 1930s collective bargaining at the insistence of the owners had been undertaken strictly on a district basis. The owners frequently refused to contemplate national negotiations and it was not until after the wages dispute of 1935 that they agreed to nominate members of a national body which might, among other things, discuss national wages questions.

During the Second World War the industry's collective bargaining and conciliation machinery was thoroughly overhauled after the publication of the report of the Greene Board. The form of machinery adopted was concerned with national discussions, national negotiations and national conciliation. District bargaining was provided for but only on a subsidiary basis. This change in the structure of the bargaining machinery, allied to the prospect of the nationalisation of the industry in the near future, undoubtedly acted as a catalyst on the miners' moves towards one national union.

The moves which finally resulted in the setting up of a national union were started in 1942. At the annual conference of that year the Federation executive was given a mandate to prepare draft proposals for the reorganisation of the union. The plans were carefully prepared by the executive. They comprised a judicious mixture of district autonomy and central control. All existing district unions were to become areas in the new union. The areas would levy their own subscriptions and have their own executive council and officials. The constitutions of the area unions, however, would have to confirm to a model drawn up by the central executive, all members would have to pay a levy (5d. was suggested) to the central organisation and if any area organisation tried to break away from the central body then the central body would have the right to organise an 'official' branch in that area.

These plans were discussed at a special conference in August 1944. A number of minor amendments were proposed and accepted. The levy was to be 4½d. not 5d., certain policy decisions (a commitment to the nationalisation of the industry, for instance) were to be included in the constitution, but the main part of the plans escaped unscathed. Two months later the proposals were submitted to a ballot of members and accepted by 430,630 votes to 39,666. Further a majority in favour of the proposals was secured in every single district. As a result, the National Union of Mineworkers was established in 1945.

Conclusion

The period of the Second World War in many ways marked a turning point in the fortunes of the coal-mining industry. During the war the industry was to leave behind the depression of the 1930s. In future there was to be a shortage, rather than a surplus of coal, and a shortage rather than a surplus of coal-miners.

The government during the Second World War was keen, as it had been between 1914 and 1918, to protect the supply of coal. Coal was essential for the maintenance of the war effort. Government policy during the Second World War, however, was hardly any more decisive than it had been twenty-five years earlier. It was essentially piecemeal and tentative. The government only took decisive action when all else had failed. Thus, attempts would be made to persuade men to return to the industry voluntarily; only when persuasion had failed would compulsion be attempted. Again, the industry would be left to organise itself for the war effort; only when it had clearly failed to do so would the government intervene. The government's performance during the war was summed up cruelly, yet fairly accurately, by *The Economist* in the following way:

> The timidity of the (Mines) Department — or, rather, the Government as a whole, since a single subsidiary Department has limited powers in a matter so charged with political dynamite — has amounted almost to cowardice. It has imposed only the loosest possible form of control; it has retreated at every growl from either side; it is almost inconceivable that it could have done less than it has.[58]

The miners for their part were clearly disappointed with government action during the war. They argued for stricter government control over the industry, for stronger action over manpower, for the rationing of coal supplies, and so on. Yet the miners were undoubtedly willing to cooperate with the government in the war effort despite their reservations about government policy. Frequent calls were made by the miners' leaders for greater productivity, less absenteeism, and so on.

The miners also clearly gained a considerable amount during the war. They were able to use the improved position of the industry to achieve a number of their long-term aims. Miners' earnings increased significantly from £3 1s. 9d. a week in 1939 to £5 16s. 11d. a week in 1945 (the working-class cost of living index increased by about 30 per cent over the same period); a minimum wage was introduced; and national negotiations were secured.

The mine-owners were obviously less happy with events in this period. They argued that the government was moving in the wrong direction. Increased government intervention was not the way to increase production; that had been proven during the First World War. Increased wages unrelated to productivity were also a backward step. They pointed to the increase in absenteeism which accompanied such increases in wages as evidence of that.

Yet, the mine-owners, like the mine-workers, did not carry their opposition to government policy to extremes. While publicly disapproving of government intervention in the industry they agreed to cooperate with the measures that were taken. The mine-owners, of course, in many ways were more concerned with what was likely to happen after the war than with what happened during the war.

The decision to nationalise the industry after the war was obviously of immense importance. Overnight it transformed a great deal of what had previously been taken for granted. No longer was it necessary to persuade the owners to amalgamate concerns, undertake programmes of modernisation and mechanisation, raise wages and the like. The government were now the owners and it was the government that would be congratulated or criticised for any action of lack of action in these matters.

The form nationalisation took is important not merely for the future administration of the industry but also because it indicated a change of view of many of the parties involved. The early claims by the mine-workers and the Labour Party for greater 'workers' control' of the industry had long since been abandoned. The miners had changed their position since the 1920s; now they argued for nationalisation through a public corporation and the maintenance of the independent position of their unions.

NOTES

1. Made under Control of Employment Act, 1939.
2. Armed Forces (Conditions of Service) Act, 1939.
3. National Service (Armed Forces) Act, 1939.
4. Fuel and Lighting Order, 1939, made under regulation 55 of the Defence Regulations, 1939.
5. *The Economist,* 24 February 1940, p.329.
6. Essential Work (Coal-mining Industry) Order 1941, made by the Minister of Labour and National Service under Regulation 58A of the Defence (General) Regulations, 1939.
7. *Op.cit.,* pp.141-2.
8. Registration for Employment Order, 1941.
9. *The Economist,* 6 September 1941, p.304.
10. *The Times,* 23 September 1941, p.5.
11. W.H. Court, *op.cit.*
12. H. Wilson, *op.cit.*
13. Board of Trade, *Memorandum on the Production, Distribution and Rationing of Coal,* Cmd. 6364, H M S O, London, 1942.
14. *Ibid.,* p.3.
15. *Ibid.,* p.3.
16. Ministry of Fuel and Power, *Recruitment of Juveniles in the Coal Mining Industry* (Chairman Foster), H M S O, London, 1942.
17. H. Wilson, *op.cit.,* pp. 53-8
18. *Op.cit.* p.4.
19. *Ibid.,* p.7.
20. *Ibid.,* p.7.
21. *Ibid.,* p.7.
22. Board of Investigation, *The Immediate Wages Issue* (Chairman Greene), H M S O, London, 1942.
23. *Ibid.*
24. *The Times,* 24 June 1942, p.2.
25. Board of Investigation, *Supplemental Report: Output Bonus* (Chairman Greene), H M S O, London, 1942.
26. Board of Investigation, *Third Report: Machinery for Determining Wages and Conditions of Employment* (Chairman Greene), H M S O, London 1943.
27. Board of Investigation, *Fourth and Final Report* (Chairman Greene), H M S O, London, 1943.
28. *The Times,* 21 July 1943, p.2.
29. *Ibid.,* p.2.
30. Announced by the Minister of Fuel and Power in the House of Commons, 12 October 1943.
31. National Tribunal for the Coal Mining Industry, *Fifth Report* (Chairman Porter), H M S O, London, 1944.
32. Quoted by H. Wilson, *op.cit.,* p.101.
33. *Financial Position of the Coal Mining Industry: Coal Charges Account,* Cmnd.6617, H M S O, London, 1945, p.5.
34. *Ibid.,* p.3.
35. Hansard Volume 397, Column 41, quoted by H. Wilson, *op.cit.* p.102.
36. *Op.cit.*
37. *Op.cit.*

38. *Op.cit.* pp.58-9.
39. Reid Report, Cmd. 6610, *op.cit.*
40. *Ibid.,* p.1.
41. *Ibid.,* p.37.
42. *Ibid.,* p.37.
43. *Ibid.,* p.138.
44. *Ibid.,* p.138.
45. R. Foot, *Plan for Coal,* Mining Association of Great Britain, London, 1945.
46. Appointed from outside the industry for the purpose of formulating the owners' plans for the industry after the war.
47. Quoted by W.A. Lee, *Thirty Years in Coal,* Mining Association of Great Britain, London, 1954, p.192.
48. Quoted by J. Platt, *British Coal: A Review of the Industry, its Organisation and Management,* Lyon, Grant and Green, London, 1968, p.19.
49. *The Times,* 14 October 1943, p.8.
50   Q. Hogg, Col. Lancaster, P. Thorneycroft (eds.), *Forward by the Right: A National Policy for Coal,* Tory Reform Committee, London, 1944.
51. Quoted in *The Times,* 4 April 1944, p.2.
52. *The Economist,* 16 June 1945, p.809.
53. Quoted by A.H. Hanson, 'Labour and the Public Corporation', *Public Administration,* Vol.32, 1954, p.205.
54. *Ibid.,* p.207.
55. *Ibid.*
56. G.D.H. Cole, *National Coal Board,* Fabian Society, London, 1949 (Revised Edition), pp.10-11.
57. *The Economist,* 25 July 1942, p.104.
58. *The Economist,* 11 April 1942, p.490.

4.   COAL AT ANY PRICE

The Coal Nationalisation Act, 1946, created a new body, the National Coal Board, to run the industry. The Board was to consist of a chairman and eight other members, all appointed by the Minister of Fuel and Power. Board members were not to be civil servants and the Board itself was to be independent from the government machine.

The Board's duties were laid down in a very general fashion by the Act. They were defined as:

(a) working and getting coal in Great Britain, to the exclusion (save as in this Act provided) of any other person;
(b) securing the efficient development of the coal-mining industry; and
(c) making supplies of coal available, of such qualities and sizes, in such quantities and at such prices, as may seem to them best calculated to further the public interest in all respects, including the avoidance of any undue or unreasonable preference or advantage.[1]

The Board itself was left to decide exactly how it should discharge these duties; what kind of organisation it would need and other arrangements.

It was, however, given specific instructions on certain issues. For example, the Board was to set up machinery for consultation with its own employees and coal consumers. In the case of coal consumers the machinery itself was specified. There were to be two councils, the Industrial Coal Consumers' Council and the Domestic Coal Consumers' Council. The members of these councils were to be selected by the Minister of Fuel and Power to represent both Board and relevant consumer interests. The councils were not to have any executive powers but were to have fairly wide powers to discuss and make recommendations to the Minister on most matters concerning the sale and marketing of coal.

One of the other major issues dealt with by the Act was that of compensation for assets nationalised. In the compensation provisons a distinction was made between the coal industry itself and subsidiary activities, such as coal ovens, retail sales and the like. Compensation for the industry itself was to be paid out of a 'global sum' which represented the estimated value of the industry. The size of the 'global sum' was determined by a tribunal, chaired by Lord Greene. It fixed the value of the industry at £164,660,000. Compensation for the

subsidiary activities was paid separately on the basis of the market value of the items concerned, considered in 'compensation units'.
Compensation was paid in the main through government stock although some cash payments were made.

The Supply of Coal and the 'Coal Crisis'

The Board, when it took over control of the industry on vesting day, 1 January 1947, inherited what was probably the country's most serious ever 'coal crisis'. Fears about the supply of coal can be traced back to early 1946; coal stocks had started to decline and parts of industry, particularly heavy industry, were living on a 'hand to mouth' basis. The position eased slightly in the summer but this proved to be only a temporary respite and by the winter of 1946 there was once again serious concern about the supply of coal. The government asked industry to use as little coal as possible; reductions in consumption of 5 per cent were suggested; and cuts of two and a half per cent were imposed on the nationalised gas and electricity industries.

By the turn of the year coal stocks had fallen to only 8½ million tons, 4 million tons less than they had been the previous winter and 6 million tons less than they had been at the end of 1939. The situation became even more serious, however, early in 1947. Existing problems were exacerbated by appalling weather conditions. The poor weather started on 23 January and continued throughout February. It affected the coal position in three main ways. First, the cold weather brought increased demand for coal, in particular from domestic consumers and the gas and electricity industries. Secondly, it made mining more difficult. Many mines were flooded and made unworkable, in addition to which, many miners were unable to reach work because of the extreme weather. Third, the weather affected the transportation of coal. The railways found considerable difficulty in moving coal and many loaded wagons accumulated in railway sidings; canals froze over; and many roads were blocked.

By February 1947, therefore, there was an extremely serious shortage of coal. The most important result was the whole disruption of industry. During February nine million workers were affected by power cuts, necessary because of the lack of coal, and about half of them were temporarily unemployed. Reporting early in 1947 *The Economist* described the widespread layoffs and short time throughout British industry.

> Austin Motors head the long list of firms temporarily stopped in the Midlands and the Rolls Royce works at Crewe have introduced a shorter working week. About 8,000 workers are affected by the closing of Cadbury's Bounrville factory for ten days . . . Several

71

thousand workers are idle, or on short time, in the cotton and wool textile industries . . . Iron and steel, chemicals, motor vehicles, railway wagons and rubber tyres are all adversely affected by lack of coal.[2]

On 7 February the government prohibited the export of coal and three days later all but essential industries in the South East, Midlands and North West were asked to stop using electricity. Domestic consumers were instructed to use no power during certain hours and the railways were told to give absolute priority to coal traffic. Restrictions were also introduced on the newspaper industry; newspapers were instructed to reduce the average size of issues by a fifth and periodicals were instructed to cease publication for at least two consecutive issues.

The situation did not begin to improve until March and even then coal stocks remained ominously low. As a result the government retained many of the restrictions on coal consumption in order to build-up stocks in an attempt to avoid another 'coal crisis' in the 1947-8 winter. Eleven per cent of coal output was earmarked for stocks. The use of gas and electricity by both domestic and industrial consumers was closely controlled. Domestic coal consumers were restricted to 34 cwt. of coal a year in the South and 50 cwt. a year in the North, while industry was restricted to two-thirds of its needs.

These restrictions plus the coal imported from USA and Poland led to an improvement in the position in mid-1947. The improvement continued throughout the rest of 1947 and into 1948. In its review of 1948 the National Coal Board was able to report

> On 1st January 1948, few traces remained of the critical situation in which the National Coal Board had taken over twelve months before: more men were working in the collieries and output was rising steadily; stocks of coal were sufficient to carry the country through to the end of the winter; there was no fuel crisis.[3]

The 'coal crisis', however, was not yet over. Although the position continued to improve in 1949 and the industry began exporting coal again, the situation changed dramatically the following year and the 'coal crisis' reappeared.

The reasons for the dramatic change in the situation are many and complex but they involve at least two main factors. The first, as with the coal crisis of 1946-47, concerned the weather. The cold weather towards the end of 1950 brought about a significant increase in coal consumption; in December the power stations had used 18 per cent more coal than they had in the previous year and domestic consumption was 1½ million tons higher. The second, and probably the most important factor, was the increase in industrial production. Industrial production had taken a while to recover from the effects of the Second World War but by 1950 many of industry's problems had been overcome. The increase in industrial production, in general economic terms, was widely welcomed, but it posed major difficulties for the coal-mining industry. Some increase in production and some rise in the demand for coal had been expected but the size of the increase in both areas took the Coal Board and the government by surprise. Between 1948 and 1949, for example, coal consumption had increased by only 1.71 per cent in the engineering industry; between 1949 and 1950 the increase was more than three times as great.

In an attempt to ensure adequate supplies of coal for power stations the government severely restricted deliveries to other customers. Domestic premises were restricted to a total of 6 cwt. of coal for the months of January and February in the South, East, and South West, and 8 cwt. in all other regions. Industry was restricted to 85 per cent of its normal needs, while cuts were made in passenger rail services (British Railways were compelled to withdraw over 3,000 passenger services accounting for about 290,000 engine miles). Many industries were hit by short time and layoffs and a number of schools were forced to close down because of the shortage of coal.

The coal crisis of 1950-51 was not, however, as serious as its predecessor of 1946-47. Further, it was to be the last major coal crisis of the post-Second World War period. This does not mean that there were no coal shortages after 1951. Between 1951 and 1956 there were frequent shortages of coal and frequent alarms that another 'coal crisis' was imminent. The shortage of coal meant a continuation of restrictions, especially on domestic consumers, and the Board was forced to import substantial amounts of foreign coal to augment its own production. In 1955, for example, Britain imported more than 11 million tons of coal, almost 5 per cent of her total consumption. The shortage of coal, however, during the 1951 to 1956 period was not critical. The shortage held back industrial expansion but it did not lead to the widespread layoffs of earlier years.

The National Coal Board's Reaction

The National Coal Board, then, between 1947 and 1956 inherited and had to deal with a situation in which coal was in short supply. The seriousness of the shortage decreased somewhat towards the end of the period but there was still a shortage in 1956, especially of certain kinds of coal. The importance of this is that it meant that the Board, throughout the 1947 to 1956 period, was faced with a constant demand and expectation that it would increase the supply of coal. Merely to maintain existing supplies would have been seen as failure.

Coal consumption rose from 190 million tons in 1947 to nearly 230 million tons in 1955 and the greater part of this increase was accounted for by industrial rather than domestic consumption (see Table 8). For example, between 1947 and 1956 consumption in the electricity industry rose by over 18 million tons and in the gas industry by over 5 million tons while domestic consumption rose by less than 1½ million tons. Industry had to expand after the devastation of the Second World War but the rate at which it could expand was to a large extent determined by the supply of coal. It was of the utmost importance, therefore, that the Coal Board took whatever steps were necessary to increase the supply of coal.

Reconstruction of the Industry

At the centre of the Board's efforts to increase the supply of coal were its plans for the reconstruction of the industry. Work began on the plans soon after the Board assumed responsibility for the industry. It was not, however, until 1950 that it was ready to present these plans for approval.

In 1950 the National Coal Board published its plan[4] for the future production of coal. Its first task was to estimate likely demand. It recognised that 'any estimate of long-term demand can be little better than an informed guess'[5] but argued that nevertheless an estimate of future demand was essential. The Board believed that coal's dominance was likely to be challenged by atomic power in the coming years. World-wide, however, it suggested that coal would continue to be in demand on a massive scale for the forseeable future.

> The fact remains that Europe alone (excluding USSR) still needs 500 million tons of coal a year and it is not likely in the forseeable future that any alternative sources of power will replace more than a fraction of this enormous total.[6]

It estimated, therefore, that the total demand for British coal in the period 1961 to 1965, for all uses (for inland supplies as well as exports) would be 240 million tons a year. This was based on an estimated demand of between 205 and 215 million tons for home consumption, made up of 56-58 million tons from consumers of carbonisation coal (gasworks and gas ovens), 138-145 million tons from consumers of steam-raising and domestic coal (railways, electricity, general industry. home coal users) and 11-12 million tons from the collieries (including miners' coal). The margin of error in the total estimate, it suggested, might be as much as 10 million tons either way.

Having determined likely demand, the Board turned to the measures that needed to be taken to ensure that this demand would be met. With

74

the help of a detailed survey of existing capacity the Board argued that in order to meet likely future demand a comprehensive programme of reconstruction and investment would be needed, leading to expenditure of £635 million in the 15-year period up to 1956 (at 1949 prices). The reconstruction envisaged included investment in 250 existing collieries, opening a score of new large collieries and 50 new draft mines, and the closure of some 350 to 400 pits.

The Board's recognition of the difficulties of long-term planning were well borne out. Within a short period of time it became clear that the 1950 plan would have to be revised. A number of factors were at work. For example, the loss of capacity due to diminishing reserves at older mines was greater than had been expected; it was estimated that none of the 280 collieries which in 1955 produced nearly 30 million tons of coal would be working in fifteen years time. There had also been a general rise in prices which had affected the Board's estimates for the cost of production.

In 1956 the Board published a review of the 1950 plan in a document 'Investing in Coal'.[7] Demand for coal in 1965 was estimated at 240 million tons with a rise of 10 million tons to 250 million tons in 1970. A substantial increase in the level of investment would be necessary if this demand was to be met. In order to complete the necessary construction of the industry investment totalling £1,350 million at current prices (compared to the estimate of £635 million at 1949 prices) would be needed. In order to carry through the reconstruction programme a new department was to be established at National Headquarters and Divisional Reconstruction Departments were to be established in Scotland and the North East.

Other Measures to Increase Output: Manpower

Long-term plans for reconstruction were clearly important if the industry was to be able to produce the amount of coal likely to be demanded in the future. The Board was also expected, however, to take action over more immediate problems in an attempt to enable the industry to meet current demands.

One of the major causes of the industry's production problems throughout the 1946 to 1956 period concerned manpower. The industry had faced a severe shortage of manpower during the Second World War and this shortage continued, except for one or two short periods, through the rest of the 1940s and into the 1950s. In 1946 when the decision was taken to nationalise the industry less than 700,000 men were employed compared to about 800,000 before the war. Furthermore the industry faced the loss of thousands more when the 'Bevin Boys' left the industry.

The Board, with the help of the government, took a number of measures to try to deal with the situation. First, it was made clear that 'Bevin Boys' would not be permitted to leave the industry until they had completed their period of national service. Second, the 'Ring Fence'[8] was established around the industry; this meant in effect that men with three months service or more could not leave the industry unless they obtained permission from the Ministry of Labour. Third, the National Union of Mineworkers and the National Coal Board reached an agreement which permitted the engagement of foreign workers in British mines. In January 1947 agreement was reached between the Coal Board, the NUM and the Minister of Labour and National Service for the recruitment of Polish ex-servicemen into the industry. Their recruitment was to be subject to three conditions. First, no Polish worker would be placed in employment without the agreement of the local NUM branch. Second, no Polish worker would be employed unless he agreed to join the miners' union. Third, in the event of redundancy in the industry, Polish workers would be the first to be dismissed. During 1947 about 6,000 Polish workers were placed in employment in the mines. Initially there was some resistance to their employment and a number of miners' lodges refused to giver permission for them to work in their areas. At one time there were about 1,000 Polish workers who had completed training but were unable to gain work because of local opposition to their employment. Gradually, however, this opposition was overcome and at the end of 1947 the Board reported:

> The Poles have proved to be excellent workers and get on well with the British mineworkers. Opposition to their employment has practically disappeared.[9]

The success of the experiment in recruiting Polish workers encouraged the Coal Board to widen its scope. Later in 1947 they reached an agreement with the miners' union and the government for the recruitment of European Volunteer Workers. The Volunteer workers came from many different parts of Europe, mainly from camps for displaced persons. Unlike the Polish workers, who were subject to a strict language test, many volunteers could speak little or no English. As a result their period of training was longer and covered English language as well as basic mining skills. Nevertheless, by the end of 1947 1,230 European Volunteers had been recruited and 688 were actually at work in the collieries. In 1948 and 1949 a further 8,600 and 2,300 European Volunteer Workers joined the industry respectively.

Partly as a result of these measures and partly because a number of ex-miners returned to the pits after demobilisation, the manpower position in the industry began to improve towards the end of the 1940s.

In 1949 over 720,000 men were employed in the industry, more than 20,000 in excess of the numbers in employment three years earlier.

Although the industry was not spared manpower problems altogether in the 1950s, the nature of the problem was different. The problem of the late 1940s had been an overall shortage in the number of men employed in the industry. In the 1950s this overall shortage disappeared. In 1952 and 1953, for example, restrictions were placed on recruitment in certain areas and the National Coal Board reported that by the end of 1952 thirty areas had as many or more men than they needed. The problem of the 1950s was one of selective rather than overall shortages. While there was sufficient manpower in the industry as a whole, and more than sufficient in one or two places, there was a shortage of manpower in certain areas. Even more crucial, generally, the areas of manpower shortage were the most productive ones. Thus severe manpower shortages were reported in Yorkshire and the Midlands.

One of the major difficulties facing attempts to remedy this situation was the change in wage bargaining which had started during the war and had been consolidated during nationalisation. It was no longer possible in the early 1950s to vary wage rates to attract workers to the districts where there was a labour shortage; as a result other methods had to be tried. In the autumn of 1950 a sub-committee of the National Consultative Committee had examined the problem of attracting and keeping manpower in the industry and suggested eight steps that might be taken in this direction:

(i)   improvements in housing in mining areas;
(ii)   the provision of more hostels for temporary accommodation;
(iii)   a better reception for new recruits to the industry;
(iv)   systematic interviewing of intending leavers to try to discover why the industry was proving 'unattractive';
(v)   stopping the call-up of army reservists engaged in mining;
(vi)   improvement of opportunities for 'up-grading' in the industry;
(vii)   improvement of amenities in colliery villages;
(viii) postponement of a proposal that had been made to lift the control that existed over the areas in which foreign workers could gain employment.

Some of the items mentioned in this report were used by the Coal Board on a selective basis to try to entice recruits to certain areas. Primary emphasis was placed on housing.

The Board believed that housing could help in two ways. First, it believed that the lack of adequate housing was preventing men working in certain areas.

Information from the coal-fields showed beyond doubt that in many places the shortage of housing was losing coal production. Sometimes men were having to travel too far to the pits; as they could not find a house near their work, they decided to seek jobs in industries near their home. Elsewhere the shortage of houses was hindering the recruitment and transfer of men to expanding collieries. In all there were about 80 districts, mostly where pits were being developed, in which to provide more houses was the only way to expand — indeed, sometimes to maintain — coal production.[10]

Second, it believed that because of the general housing shortage the offer of housing to recruits could be an important incentive to attract manpower into the industry.

In 1952 the government agreed to increase the allocation of houses for miners: 20,000 extra houses were to be built in areas where there was a shortage of labour; 9,000 in Yorkshire, 8,500 in the Midlands, and the rest in selected parts of Lancashire and South Wales. In an attempt to expedite the construction of these houses the National Coal Board set up its own housing association in the spring of 1952. The Association worked with local authorities in the areas concerned to produce housing estates that would contain a mixture of miners' and non-miners' houses. By the end of 1953 over 5,000 miners' houses had been completed and the bulk of the rest of the special allocation was completed in the following year.

Undoubtedly moves such as these helped to improve the position somewhat in the worst areas. They did not, however, entirely solve the problem; difficulties were still being encountered in 1955 in certain areas. Manpower in the North Eastern, West Midlands and South Western Divisions had fallen by 6,400 during 1955. At the end of 1955 it was estimated that the industry would still need about 13,000 more men, and 9,000 of this total was required for the three mentioned Divisions.

The level of manpower in the industry, though, was only one of the 'manpower' problems facing the Board in the late 1940s and early 1950s. The Board was also concerned to increase the number of shifts worked by the manpower available.

When it took over responsibility for the industry on 1 January 1947 the Board was faced with an understanding between the government and the miners that the length of the working week in the industry would be reduced; a five-day week would be introduced as soon as possible. The Board immediately began discussions with the unions over the issue and in April 1947 an agreement was reached and the five-day week was introduced the following month. Later in 1947, because of the need to increase coal production, an agreement was concluded between the Board and the unions for an extension of working hours. The extension of working hours was to take the form of Saturday working or an addition to the normal working day. The method adopted was left to individual areas to determine and any such extension of working hours was to be paid at overtime rates. In fact, most areas agreed to Saturday working although an addition to the normal working day was decided upon in Northumberland and parts of Durham.

These changes in the hours of work, however, were in many ways only of marginal importance to the industry. This was largely because of the industry's long-standing problem of absenteeism. To a certain extent the length of the official working week did not matter; what mattered was the average number of shifts actually worked by the miner.

It is, though, easier to describe the nature of the problem than to explain why it existed or what might have been done about it. Almost all of the investigations (and there have been many) into the subject have produced a different cause and a different cure for absenteeism in the industry.

A study by Political and Economic Planning, [11] for example, suggested that there was some evidence to support at least a modified version of the contention put forward by the coal-owners during the war that high wages were the cause of absenteeism.

A graph published in the *Daily Telegraph* of December 4, 1945, traced the correlation between absenteeism, wage rates, and the price of coal, and the conclusion was drawn that the price of coal was high because absenteeism was high, and that absenteeism was due mainly to high wages. It is certainly true that higher wages permit the miner to extend his period of absence when ill or lengthen his holiday, but they hardly explain why, in fact, he should do so any more than workers in other industries. The rise in wage rates in the coal-mining industry has, however, been more rapid than in almost all other industries, and there may be some truth in the contention that many miners, having become accustomed to a relatively low standard of living, were not particularly interested in increasing their earnings beyond a certain level.[12]

Alexander, however, in a later study, [13] challenged the relationship between wages and absenteeism. He examined in detail changes in wage rates since nationalisation and accompanying changes in absenteeism. Of the five major wage increases looked at two were followed by decreases in absenteeism totalling 4½ per cent while three were followed by increases in absenteeism totalling 3 per cent. He also examined variations in absenteeism levels between different districts and different pits and said:

> On the evidence here collected it is possible to conclude that wage increases do not appear to increase absenteeism, that there is a very slight tendency for high wage districts to have higher than average absenteeism but that a closer inspection at pit level does not endorse the view that there is a clear positive relationship between earnings and absenteeism.[14]

Baldwin, [15] on the other hand, argued that the problems of absenteeism were being obscured by studies that saw absenteeism as a general phenomenon. Absenteeism was only high amongst a small minority of workers. He quoted a survey by the Board's East Midlands Division in February 1950 which found that 6 per cent of the men employed accounted for 49 per cent of all absences. The survey also went some of the ways towards identifying the high absentees. Men in their early twenties and faceworkers tended to take most time off, and absenteeism was high amongst men who worked night shifts. Baldwin concluded from this evidence:

> There can be no doubt that almost all men take an occasional unexcused absence ... But there seems equally little doubt that the 'problem' of absenteeism is primarily one created by a fairly small minority of men who take off an undue amount of time.[16]

Similar conclusions were reached by Liddell in a study of absenteeism in the Cannock Chase and Durham coal-fields. [17] In Cannock Chase it was noted that the high rate of absenteeism amongst faceworkers was accounted for by a small proportion of the men, probably about 10 per cent, while in Durham it was found that in one colliery about 13 per cent of the men accounted for over one-third of the absences, whereas 85 men, mainly surface workers, were absent for less than two days per year, and virtually all of this absence was the result of sickness.

Although both the Board the government recognised that absenteeism was likely, because of the nature of the work, to be higher amongst coal-miners than other workers and although neither was sure that they new how to reduce the level of absenteeism, both felt that it was necessary to make an attempt to do so. Broadly, two types of measure were taken in this direction. First, financial incentives were tried to improve attendance. Thus, when the five-day week was introduced it was agreed that the full wage would only be paid if there was full attendance for the week. Effectively the pay from the sixth day of the

80

old working week week was used as an attendance bonus. In fact, this move had only limited success. One of the reasons why it was not more successful was that the list of acceptable excuses for non-attendance became so long that very few absences were actually penalised. Second, penalties for persistent absentees were used. During the war absentees had been prosecuted but with the end hostilities the support for such measures faded. In 1947 the National Coal Board asked the Colliery Consultative Committees to help to fight absenteeism by interviewing persistent absentees and suggesting what action might be taken against them. In its report of 1947 the Board recorded the type of action taken by the Consultative Committees:

> In the North Eastern Division habitual absentees are asked to attend the colliery meetings to explain themselves. In the South Eastern Division, the consultative committees have suggested the dismissal of habitual absentees. In the East Midlands Division during the year two hundred men were dismissed for persistent absenteeism on the advice of the Colliery Committees.[18]

These measures, though, seem to have had relatively little effect. Table 9 shows absenteeism rates between 1946 and 1956. It can be seen from this table that absenteeism declined immediately after the war; from 16·4 per cent in 1946 to 12·5 per cent in 1947. This fall, however, is largely accounted for by the introduction of the five-day week. As a result, although absenteeism dropped, so did attendance, from an average 4·86 to 4·72 shifts over the same period. For the rest of the post-war years absenteeism rates, despite the measures outlined, stayed fairly constant at about 12 per cent.

Other Measures to Increase Output: Organisation and Administration of the Industry

The manpower problems that faced the industry in the 1946 to 1956 period were obviously of tremendous importance. Many of the Board's critics argued, however, that the Board could have made better use of the manpower when and while it was available. Attention centred on productivity and the organisation and administration of the industry.

Between 1946 and 1956 output per manshift (see Table 10) increased from 1·03 tons (2·76 at the coal face) to 1·23 tons (3·33 at the coal face). It was suggested that given the extent of the investment in the industry (between 1947 and 1956 the Board's capital expenditure totalled well over £500 million) the increase in productivity was disappointing. Thus:

> When the expanding volume of investment under nationalisation

is considered the small rise in productivity is disappointing.[19]

Many of the Board's critics argued that the rise in productivity had been so disappointing because of the poor organisation and administration of the industry. One of the Board's fiercest critics in this area, Sir Charles Reid[20] argued that the industry could have produced 30 million tons more coal, with the same machinery and the same manpower, if only its organisation had been better.

The Board had, in fact, been given a fairly free hand by the Nationalisation Act to select the kind of organisation it felt would be most suitable for the industry's needs. The Board decided to base the industry's organisation on a series of Divisions and Areas. Initially eight Divisions were created:

    (i)   Scottish Division — Scottish Coal-fields;
    (ii)   Northern Division — Northumberland, Durham and Cumberland Coal-fields;
    (iii)   North Western Division — Lancashire and North Wales Coal-fields;
    (iv)   North Eastern Division — West and South Yorkshire Coal-fields;
    (v)   West Midlands Division — N.Staffordshire, S.Staffordshire, Cannock Chase, Warwickshire and Shropshire Coal-fields;
    (vi)   East Midlands Division — Nottinghamshire, Derbyshire and Leicestershire Coal-fields;
    (vii)   South Western Division — South Wales, Forest of Dean, Bristol and Somerset Coal-fields;
    (viii)   South Eastern Division — Kent Coal-field.

In 1950 the number of Divisions was increased to nine by the splitting of the Northern Division into two parts, one called the Durham Division and the other the Northern (Northumberland and Cumberland) Division.

The role of the Division under nationalisation has been variously described as a 'buffer' between the industry's national and local organisation, a 'co-ordinator' and a 'controller'. In fact, the exact role played by the Division varied from place to place but it usually covered some of the following functions: the appointment of 'area' managers, centralised marketing, provision of scientific services, control of labour relations in the Divisions, control of production (in particular control over colliery reorganisation schemes) and the provision of financial services. The Division itself was controlled by a Divisional Board, appointed by the Central Board, and consisting of a chairman (normally from outside the industry), a deputy chairman and four other members.

Each of the directors, other than the chairman and deputy chairman, was given charge of one area of the Board's work, either finance, marketing, production or labour.

Below the level of the Division the industry was divided into forty-eight areas (one Division, the South East, was so small that it was not divided into areas). The nature of the area organisation varied from place to place and changed over time. For example, most areas were organised on the basis of departments that reflected the Division and National organisations, but in some cases departments would be combined, such as labour and welfare, whereas in others they would not. Further, the Area Management Committee (usually comprised of Departmental heads) would in some areas have a great deal of executive authority, while in others it would merely advise the Area General Manager. Over the first few years the Area General Manager's position itself changed a good deal. Initially the production manager assumed the Area General Manager's role but towards the end of 1947 the two functions were split.

Although the areas had control of the collieries under their jurisdiction it was usually control of a supervisory rather than a day-to-day kind. The day-to-day supervision of the mines was arranged through yet another tier in the organisation, the sub-area or group. Under this arrangement a small number of collieries would come under the control of either an 'Agent', a 'Sub-Area Manager' or a 'Deputy Production Manager'. It was also normal for the sub-area or group to have a small selection of other officer, such as labour officer, a training officer, an engineer, a surveyor and so on, although practice varied considerably on this matter. Beneath the sub-area or group would normally be the colliery, under the control of a colliery manager, although in some cases there were special arrangements for extra administrative units between the sub-area and the individual colliery, especially where the number of collieries controlled by the sub-area was large.

The criticisms of the industry's organisation and administration might be looked at under two headings; the issue of 'functional' Board members and 'decentralisation'. Each member of the National Coal Board, apart from the Chairman and Deputy Chairman, had in 1947 been given responsibility for a specific sphere of the Board's work. This method of organisation was strongly criticised by one of the Board's most influential members, Sir Charles Reid (Chairman of the Committee that produced the 1945 Technical Advisory Report). Reid, who resigned his membership of the NCB in 1948, partly as a result of this issue, argued[21] that because the Board was composed of functional members, these members were unable to take a detached view of the Board's work. They were, in essence, too concerned with their own departments and their own responsibilities to be able to take a broader view of the industry. He suggested[22] the reconstruction of the National Board so that it consisted of a Chairman, three Vice-Chairmen (all full-time members) and eight part-time members. None of the members of the Central Board would have specific departmental responsibilities.

This point of view was given some support by the report of the Burrows Committee. The Burrows Committee had been appointed by the National Board in May 1948 to look into the industry's structure and administration and consisted of three members, one of whom, Sir Robert Burrows (a former prominent coal owner who later became the Board member for marketing) was chairman. The Burrows Committee made its report to the Board later in 1948 and some of its recommendations (although much to the disquiet of contemporary commentators, not all) were published by the Board in November 1948. The Committee did not recommend the abolition of all of the functional responsibilities of Central Board members but it did recommend the delegation of many of those responsibilities to chief officials and the appointment of a number of extra members without such responsibilities (it also made a number of recommendations on other matters, such as the frequency of meetings of the Central Board and the organisation of the industry below the national level).

The Board accepted many of the Burrows Committee's recommendations. The Coal Industry Act of 1949 included a provision, at the request of the Board, which allowed the Minister to appoint up to twelve members to serve on the Central Board of whom three might be part-time. A second deputy chairman and a number of part-time members were appointed without functional responsibilities.

It is interesting to note, however, that the Board subsequently reversed part, and in some ways the most important part, of the action taken on the Burrows Committee report. Late in 1953 the Board appointed yet another committee to look at its organisation. This committee, under the chairmanship of Dr Fleck (at the time Chairman of ICI) reported in February 1955. Its suggestions ranged from a reorganisation of the personnel function (it should be divided between two new departments, Industrial Relations and Staff) to limits on the Area General Manager's powers and other changes in lower level organisation. On the question of functional membership of the Central Board, the Committee recommended an increase in the functional responsibilities of members; all the full-time members of the Central Board, except for the Chairman and Deputy Chairman, should have departmental responsibilities. These recommendations were broadly accepted and put into effect by the Board.

The debate over the centralisation of the Board's activities was another to which Sir Charles Reid contributed.[23] He argued that the coal industry was over-centralised. All major decisions were taken by the Central Board and the Divisions; no scope was left for the Areas, the most crucial units of production

> The areas, of which there are 49 in the country, are the real units of production where the actual management of the mines is done. The general managers, who are the mining experts there, find themselves deprived of the power to exercise initiative and are often frustrated by the decisions of the divisional boards, in which they have no real confidence. They do not feel themselves to be in charge of operations as they were under private enterprise. They are unable to make decisions which ought to be within their province, and their status and authority in the eyes of the staff and workmen are reduced.[24]

He suggested that the industry's organisation should be based on the areas. The number of areas should be reduced to twenty-six and they should be made into separate corporations responsible for all activities under their control. The area itself should be under the control of a Managing Director, who should be supported by a Board consisting of certain departmental heads, possibly four part-time members, and himself. One of the part-time members should act as chairman. Under this scheme the Central Coal Board would be reduced drastically, both in size and importance. Apart from the Chairman and Vice-Chairman, all other members would be appointed on a part-time basis and the Board would merely operate as a holding company.

Support for Reid's views came from a number of directions. The Conservative Party's criticisms of the industry stressed the disadvantages of centralisation and their plans had a great deal in common with those put forward by Reid. Outlining these plans Butler stated that

> some twenty to thirty autonomous area boards should be set up in which full executive control would rest with their general managers, that the divisional boards should be abolished and that the functions of the National Coal Board at the centre should be confined in scope to the questions of the raising of capital, major financial development plans, such questions as national wage negotiation, co-ordination of selling prices and such common services as research.[25]

These plans only differed from those of Reid in one major respect; although they envisaged a significant reduction in the powers of the Central Board they did not go so far as to suggest that it should merely act as a holding company.

Similarly, Burn, in an article[26] on the National Coal Board argued that although the benefits of decentralisation were unlikely to be as widespread as Reid suggested, they were nevertheless likely to be substantial and worthwhile.

The support for Reid's views, though, was by no means universal. Chester, writing for the Acton Society Trust, argued that the criticisms of Reid and the Conservative Party were based on a misunderstanding of the situation. Centralisation in the industry was a problem but centralisation was most evident at the area, not the national level. Consequently, the plans of both Reid and the Conservative Party, which would divest the National Board of many of its responsibilities would exacerbate rather than ease existing problems.

> The principle defect of proposals like these is that they provide measures for splitting the industry into a number of units, each of which itself may remain highly centralised and each of which is large enough for such centralisation to be felt as oppressive by operational management. They achieve decentralisation at the uppermost level, but not at the lower levels. But it is precisely at the uttermost level where the degree of decentralisation is already greatest . . . Such proposals seem to be based on a mistaken analysis of the problems.[27]

The National Coal Board's reaction to criticisms over centralisation ran along two lines. First, they claimed that some degree of decentralisation had already taken place. It is, of course, 'difficult to cite objective information on functions which have been decentralised; decentralisation is more a matter of spirit which eludes expression in concrete form.'[28] There is, however, a statement from the Chairman of the Board, Lord Hindley, in 1947,[29] about attempts to decentralise the organisation and the evidence of an increase in certain of the powers of the areas, to support the idea that there were some moves in this direction. It is clear, though, that the Board also attacked the criticisms by stating that it felt it would be wrong to take decentralisation too far. A number of arguments were put forward in support of such a philosophy. First it was the Board's duty to oversee activities in the 'national interest'; if matters were left merely to the Divisions or the Areas then they would be left to 'local monopolies' who could not be expected to take as much notice of the 'national interest' as the National Board. Second, if the Divisions or Areas were given complete autonomy then because many of them were operating at a loss they would be forced to increase the price of coal. Current arrangements meant that the losses of one Division or Area were balanced out by the National Board against the profits of another Division or Area. Third, if the Areas or Divisions were given greater autonomy this would lead to greater political interference in the industry's affairs. The government would, because of the greater independence of these smaller units, have to deal directly with them.

The Role of the Government

Under the terms of the Nationalisation Act control of the coal-mining industry had been given to an independent body rather than to a government department. The intention had been to separate the industry from the government machine and to insulate it from direct day-to-day Parliamentary scrutiny. This did not mean, of course, that the government and Parliament had no interest whatsoever in the industry's affairs. Even under private ownership the industry had been the subject of frequent government activity and Parliamentary debate. Similarly, after nationalisation, the industry had to fit into the government's general economic plans for the country, as debated and approved by Parliament. In this context not only was the amount of coal produced crucial but so were the wages paid to miners and the price charged for coal. The government in the post-war period campaigned to keep general wages and prices under control and clearly saw the policy of the coal-mining industry as being important in this respect. The coal-mining industry, after all, employed about 700,000 people or about 5 per cent of the country's labour force and the cost of coal was a major determinant of the cost of producing a whole range of other goods. The government and Parliament after nationalisation, however, were not merely interested in general matters such as these; they also took, mainly through the Nationalisation Act, a whole range of powers to influence and control the direction of industry.

The government, for example, through the Minister of Fuel and Power, was responsible for appointing the members of the National Coal Board. Under the terms of the Nationalisation Act the government also, again through the Minister of Fuel and Power, had the right to

> give to the Board directions of a general character as to the exercise and performance by the Board of their functions in relation to matters appearing to the Minister to affect the national interest.[30]

and the Board was instructed to

> afford the Minister facilities for obtaining information with respect to the property and activities of the Board, and shall furnish him with returns, accounts and other information with respect thereto and afford him facilities for the verification of information furnished, in such a manner and at such times as he may require.[31]

87

Further, the consumers' councils set up under the Nationalisation Act were to report direct to the Minister and the Minister, if he wished, could order the Board to take action over any of the matters referred to be the councils.

These general powers to shape the industry's future were frequently used throughout the 1940s and the 1950s, although it is difficult to determine the extent to which the government forced the Board to take action against their will. The government, for example, insisted on examining and confirming the Board's plans, issued in 1950 and 1956, for the reconstruction of the industry. The government also undertook its own enquiries into the future direction of the industry. In 1952, for instance, the report of a government committee on national fuel and power resources, [32] under the chairmanship of Lord Ridley, assessed the likely future demand for coal. The committee argued that there was an acute shortage of coal in the UK:

> The United Kingdom at the present time remains acutely short of coal – supplies are insufficient to meet all demands at current prices. The shortage is masked by the 'rationing' of house coal, the allocation of coal to industry, and the existence of price control. Domestic consumers cannot buy as much coal as they would like to purchase, industrial users do not always receive the qualities of coal they would prefer, and coal exports are restricted to a level far below what could profitably be sold abroad.[33]

It estimated that the real demand for coal was about 236-241 million tons a year, about 15-20 million tons above the current level of production. It also estimated the likely demand in ten years time (1959-1963) at between 257 and 267 million tons a year (of which 25-35 million tons would be exported). In essence, however, the committee's report did little more than confirm the coal industry's own predictions. It suggested that the Coal Board's estimates were about right, although if anything they erred on the conservative side.

The government also regularly issued production targets for the industry through the *Economic Survey*. The *Survey*'s aim was to

> review the economic state of the nation . . . and to set out some of the prospects and targets, in the light of which the government intends to plan during the year.[34]

In 1947 the survey suggested coal production of 200 million tons as the appropriate target, in 1948 211 million tons, in 1949 215-220 million tons, and so on. It is doubtful, though, if these targets told the industry much more than they already knew and the government admitted that the survey was not an attempt 'to forecast events precisely, to lay down rigidly detailed plans'.[35]

The government's powers in the financial sphere were rather more important; the government had wide powers to control the Board's financial activities. For example, advances to the Board were made through the Minister of Fuel and Power, and the Board's borrowing powers were limited by legislation. Any major increase in capital expenditure, therefore, had to be sanctioned by the government. Thus, in 1951 the government passed the Coal Industry Act which enabled the Board to borrow up to £40 million in one year and up to £300 million in total in order to finance its reconstruction plans. In 1956 a further Coal Industry Act had to be introduced to permit the Board to finance its new reconstruction plans; this time the Board was given permission to borrow up to £650 million in total.

The government also exercised a degree of control over the industry's finances through pricing policy. The price of coal had been more or less under government control since 1939. Just before the beginning of the war the Central Council of Colliery Owners gave the Mines Department an assurance that the general level of coal prices would not be increased without the government's agreement. In 1941 this assurance was extended to cover the prices of different qualities of coal as well as the general price level. Before vesting day the Board renewed this assurance to the Minister of Fuel and Power. Consequently the government was able to effectively control the price of coal after the war. On a number of occasions they used this power to reduce the price increases suggested by the Board, as for example in 1947, and this had an important impact on the Board's ability to meet its obligation to generally balance revenue and expenditure over a period of years. The Nationalisation Act had instructed that

> the revenues of the Board shall not be less than sufficient for meeting all their outgoings, properly chargeable to revenue account . . . on an average of good and bad years.[36]

The relationship of the National Coal Board to Parliament, rather than to the Minister of Fuel and Power, was not spelled out in the Nationalisation Act but the responsibilities and duties given to the Minister obviously meant that Parliament would take an interest in its activities. The Board's activities were placed under the general control of the Minister of Fuel and Power and constitutionally the Minister was under the control of Parliament. Initially, Parliament took a great interest in the activities of the National Coal Board and the Minister dealt with many points in detail in defending the Board. Towards the end of 1947, however, it was decided that Parliamentary supervision was becoming too detailed and a policy decision was reached which meant that in the future the Minister would only deal with general issues and Members of Parliament would not be permitted to ask detailed questions about the day-to-day running of the industry.

Members of Parliament have also had the opportunity to discuss and if they thought fit to criticise and recommend action on, the industry's affairs on other occasions. When the government introduced legislation concerning the industry, as with the Coal Industry Acts of 1951 and 1956, this gave MPs an opportunity to enter into debate, and the government undertook to present the Annual Reports of the Board to Parliament, giving a further opportunity for discussion. Apart from this many MPs have contacted the Board directly about issues of interest to their constituents and they were given an assurance by the government that the Board would give them a sympathetic hearing.

The Reaction of the Miners and their Unions

The relationship of the Board to the trade unions in the industry had, like that between the Board and Parliament, been the subject of much debate in the period before 1945. Under the terms of the Nationalisation Act the trade unions were to have no direct relationship with or control over the Board. Both the trade unions and the Board were to retain their independence in this respect. Ebby Edwards, a former General Secretary of the Miner's Federation, joined the National Coal Board as its member for labour relations but when he did so he resigned his union membership. Formal contact between the Board and the union was reserved for the traditional conciliation and consultation machinery. The conciliation machinery was broadly that inherited from the Greene Board, of a National Negotiating Committee and a National Reference Tribunal. An important addition was made to the machinery, however, in 1946. Under the terms of an agreement between the NUM and the NCB (elect) the district conciliation scheme was extended and a new pit conciliation scheme was introduced. The pit conciliation scheme laid down six steps for the handling of pit disputes:

1. The issue would be discussed by the workman or workmen involved and the pit official.
2. If no settlement was thereby reached the issue would be discussed by the workman or workmen and the colliery manager and his representative.
3. If no settlement was reached in these discussions then the matter would be referred to the union officials and the colliery manager.
4. The next stage was for the issue to be discussed by a metting of representatives of management and the union.

5. If there was still no settlement then the matter was to be referred within fourteen days, to the Joint Secretaries of the District Conciliation Board who would arrange for the matter to be looked at by a small committee of union and Coal Board officials from outside the pit concerned.
6. Finally, the matter could be referred to an umpire chosen, by various methods, from a list compiled by the two sides. His decision would be binding, except that if he considered the matter to be of sufficient importance he could refer it to the District Conciliation machinery.

The 1946 Act specifically directed the National Coal Board to enter into consultations with its employees on

(i) questions relating to the safety health or welfare of such persons;
(ii) the organisation and the conduct of the operations in which such persons are employed and other matters of mutual interest to the Board and such persons arising out of the exercise and performance by the Board of their function.[37]

The Act did not, however, lay down the nature of the machinery to be used and this was agreed between the Board and the Unions. The consultative machinery introduced in the industry in 1947 consisted of a variety of committees at national, district, area and pit level. The national committee had twenty-seven members of whom six were appointed by the Board, nine by the NUM, nine by the industry's managerial union, the National Association of Colliery Managers, and three by the industry's supervisor's union, the National Association of Colliery, Overmen, Deputies and Shotfirers. The chairman of the Coal Board chaired the committee. The only major constraint on the activity of the committee was that it was not permitted to discuss matters which were the province of the industry's conciliation machinery, such as wages or terms of employment. The committee could deal, however, with matters such as recruitment, production, training, welfare and unofficial strikes. The district, area, and pit committees, despite minor variations, in essence mirrored the national machinery. One point, though, might be stressed about the working of this machinery. At first, there was a division at the pit level between the union and the workers' representatives on the conciliation machinery. The workers' representatives had to be members of the union but were not chosen directly by the union. This system had both advantages and disadvantages. The major advantage was that it helped to strengthen the distinction between conciliation and consultation; the major disadvantage was that it raised the possibility of an alternative source of workers' allegiance to the union. In 1948, the

91

position was changed and it was agreed that in future representatives on pit consultative committees would be chosen from nominations submitted by the union (the union was supposed to submit at least two nominations for each vacancy, although this practice was not adhered to in all instances).

Although the NUM did not have any direct say in the policy adopted by the National Coal Board it clearly influenced and supported it. The relations between the NUM and the NCB, in particular in the early years after nationalisation, were extremely good. The NUM gave a great deal of support to the NCB's efforts to deal with problems arising from the shortage of coal and it is doubtful whether the Coal Board could have achieved the degree of success it did without the union's help.

The union's support for the NCB was undoubtedly largely a reflection of its support for the idea of nationalisation. The miners' union had pressed the case for nationalisation for about fifty years and was determined to do all it could to make sure that it worked and was seen to work. The miners' union was also encouraged to support NCB policy by the government, a Labour Government which the miners had strongly supported in 1945 and for which they were willing to sacrifice a good deal.

It is also clear, however, that the union's support for the NCB was not a completely selfless affair. The union undoubtedly felt, probably correctly, that they could, providing they maintained good relations with the NCB, influence its policy and actions. There has been considerable debate and controversy over the amount of influence the NUM was able to exert over the NCB;[38] a number of the NUM's production officials resigned in the late 1940s because they claimed that the NUM was enjoying an unreasonable and unhelpful degree of influence over Coal Board policy.

Although it is probably true that in its first few years the Coal Board dealt more sympathetically with the NUM than some expected, it is also true that to a certain extent the Coal Board's reaction to the NUM was not merely based on 'loyalty', 'political sympathy' or any other such factor, but on the need to attract and retain manpower if it was to satisfactorily meet the demand for coal in the late 1940s and early 1950s.

Probably the best example of the way in which this worked can be seen by looking at the case of wages and earnings. Between 1947 and 1956 miners' average earnings increased markedly, far more than those of workers in other industries (see Table 11). For example, in 1947 the average earnings of coal-miners were 147s.1d. a week but by 1956 they had risen to 299s.1d. a week, an increase of approximately 103 per cent. Average wage rates in all industries and services only increased by 85 per cent over the same period.

This increase in the miners' average earnings was the result of a fairly steady increase in wage rates in the industry over this period. In 1947 an agreement was reached to increase the national minimum weekly wage by 15s. for underground workers (from £5 to £5 15s.) and by 10s. for surface workers (from £4 10s. to £5); general shift increases for low-paid workers (less than £6 15s. a week for underground workers and £6 5s. a week for surface workers) of 2s.6d. for underground workers and 1s.8d. for surface workers were also included in the package. In 1949 the minimum weekly wage was further increased to £6 a week for underground and £5 5s. a week for surface workers. Minimum rates were again raised in 1951 in a package costing the NCB £10 million, and later general increases of 11 per cent for underground and 10 per cent for surface workers were negotiated. Perhaps one of the most important moves, however, came in 1955 with a new agreement for the industry's day-wagemen. Following lengthy negotiations, the industry's 400,000 day-wagemen were grouped into 360 job titles and 13 different grades. Each of the 13 new grades was given a national minimum wage, in most cases a significant increase on the old rates.

There is considerable evidence that the increase in miners' earnings over this period was as much if not more the result of the industry's manpower problems than the NUM's close relations with the NCB. The rise in wage levels negotiated in 1951, for example, was specifically linked to the industry's manpower problems and followed an intervention by the Prime Minister who was concerned about the effect the shortage of manpower in the industry might have on the production of coal. Haynes, in fact, has argued that not only were the wage rises of the 1940s and the early 1950s not the result of the close relationship between the NUM and the NCB, but also that the rise in miners' earnings would probably have been greater had the industry been under private control.

Did nationalisation bring these increases in wages or would they have come anyway under private ownership? It is not at all clear that wages would have been less under private ownership – they might have been greater. The whole period from 1947 to 1951 was one of serious coal shortages, despite rising coal prices. A large part of the privately owned industry would have found it profitable to increase production, or attempt to do so. Firms would have bid against each other for the scarce labour. Indeed the various forms of 'wage freeze' and 'wage restraint' that slowed down grants of increases under public ownership would very likely have been less effective under private ownership, for the miners would probably have been more rebellious and demanding without their long-cherished goal of nationalisation.[39]

It would be a mistake, however, merely to concentrate on the reactions of the miners' leaders to the National Coal Board and their problems in the 1946-56 period. For while the miners' leaders at the national level gave their clear support to the Board, the reaction of the officials of the miners' union below the national level, and of individual miners, was rather more mixed.

Almost all miners had supported the idea of nationalisation in 1946. It seems probable, however, that they expected nationalisation to have a more dramatic effect on their working lives than actually occurred. Despite notices placed outside each mine declaring

> This colliery belongs to the National Coal Board and is Managed on Behalf of the people

nationalisation initially changed little about the miners' working day. Thus, it has been noted that

> On a day that was supposed to revolutionise his life, he (the miner) left his home at the usual hour, walked or took a bus to his pit in the usual way, went through the usual routine of putting on his pit clothes and descended into the mine. He found that the same deputy was in charge and above him the same overman, under-manager manager and manager. His pay was the same (for the first few months) and his daily task no less.[40]

Reactions, of course, varied. In Durham, Northumberland and the Midlands, the Board retained a lot of good will amongst individual miners for some while, but in other areas, of which Scotland is probably the best example, the miners quickly showed their opposition.

There is little doubt that despite the official opposition of the NUM to the idea many miners believed that nationalisation would result in greater worker participation in the running of the industry, if not 'workers' control'. Abe Moffat, the Scottish miners' leader, complained, for example, that there were too many (a majority) people on the National Coal Board without direct experience in the industry. He felt 'that there should have been more representation from the trade union movement, not a majority of people who held conservative opinions'.[41] Haynes estimated that by the early 1950s about half of Britain's miners supported the idea of more effective workers' participation in the control of their industry, and in certain areas, notably South Wales, the proportion was higher.

It has been argued, however, that a great deal of the antagonism between the Board and the individual miner could have been overcome, without any basic change in structure, if the Board had made a greater attempt to involve and inform the miners about activity in the industry.

Such views were expressed, for example, by M. Cole in a Fabian Society pamphlet on the industry.[42] She reported on the results of a survey of people supposedly 'knowledgeable' about the affairs of the industry. In this survey she found a gread deal of ignorance about the industry, especially among miners and ex-miners. She suggested a variety of measures to improve matters including changes to the consultative machinery (too many consultative committees were merely 'talking shops').

A similar view was put forward in a paper prepared by the Acton Society Trust.[43] This reported the results of another survey. The survey, the paper said, showed that

> as a general rule, it can be said that whatever falls within the miners' experience — the wages system as it affects his actual pay, conditions in the pit, local issues such as housing and transport — was known in great detail. The life of the community is built around the pit, and events touching the pit form the subject of practically all conversation. On the other hand, the wider issues and underlying causes — the economic reasons for importing American coal, the probable forthcoming shortage of manpower, the character of the National Plan, the functions of the administrative and technical machine — were largely unknown.[44]

The paper argued that the Coal Board needed to take steps to improve the miner's general knowledge about the industry and dispel some of his suspicions. The Board needed to present information to the miner more regularly and in a more simple and direct fashion.

The disparity between the views of the national officials of the NUM and the ordinary miner about nationalisation and the amount of support that should be given to the NCB inevitably caused problems. Throughout the 1946 to 1956 period the NUM's national officials concentrated on trying to persuade the miners to accept the Board's proposals and to 'make nationalisation work'. The NUM at national level consistently opposed industrial action and one of the union's area organisations (Durham) suggested that miners' lodges should reimburse the Coal Board for any losses resulting from unofficial strikes. Many miners felt betrayed by this attitude. The union was not, in their eyes, performing its traditional function of defending the interests of its own members. All too often they felt that they could not depend on the support of the union in a grievance with management. One of the results of this kind of feeling was that throughout the 1946 to 1956 period the industry was subject to a series of relatively short, small-scale, unofficial strikes.[45] Table 12 shows that before the war the industry lost more man-days through stoppages than it did after, but the position was reversed in the case of the number of stoppages. The unofficial strikes, although relatively small-scale and short, led to a considerable loss of coal production; in 1947 over 1½ million tons of coal was lost through disputes in the industry, and a similar total was reached in 1949 and 1952.

It is important to note, however, that even in the case of strike action the miners' reaction was not uniform throughout the country. Three coal-fields, Scotland, Yorkshire and South Wales, were particularly strike-prone, while other areas, like the West Midlands, Northumberland, Cumberland and Durham were relatively strike-free (see Table 13). In fact, one might be even more specific. Strike action was largely concentrated in certain areas within these coal-fields. Thus, ten areas within the three most strike-prone coal-fields accounted for over 50 per cent of the coal lost through disputes in the industry.

Conclusion

The first signs of a coal shortage in Britain appeared during the Second World War. It was not, however, until the winter of 1946-47 that the real impact of the shortage was felt. The wholesale disruption of industry that resulted from the shortage of coal left a considerable impression . The policy of the government and the Board for the next ten years was to be to try everything possible to increase the supply of coal in order to avoid a recurrence of a 'coal crisis' on the scale of 1946-47. To a large extent they succeeded. The position improved considerably in 1948 and 1949, and although the shortage of coal in 1950-51 was to cause serious problems, they were not of the same magnitude as those experienced in 1946-47.

The NCB between 1946 and 1956 took action on a fairly wide front. In 1950 it published its long-term plans for the reconstruction of the industry which were based on an estimated demand for coal of 240 million tons of coal a year between 1961 and 1965, and although these plans had to be revised in 1956, the revision merely confirmed the assumption of expansion. The NCB thus initiated a massive programme of investment to prepare the industry to meet the future demand for coal.

The NCB also took a number of measures designed to deal with the 'short-term' demands placed upon it. Such measures included efforts to increase the level of manpower in the industry, to improve the distribution of manpower between different districts, to make the best possible use of the manpower available, and to improve the industry's administrative and organisational structure.

The government's role in the industry after nationalisation is not an easy one to pin down. The government did not assume direct day-to-day control over the industry but relied on a more general overview of the industry's progress. On certain issues it seems that the views of the NCB and those of the government may have come into conflict. The NCB, for example, was more concerned to raise the level of coal prices than the government was to allow them to do so. On such matters the government was usually able to ensure that its own view prevailed.

In many ways, however, the disagreement between the NCB and the government during this period was over detail rather than more fundamental questions. On the general trend and the industry's future, the NCB and the government were broadly in agreement. Both agreed that raising the level of output should be the prime aim and the government's pronouncements on this issue (through the *Economic Survey,* The Ridley Report and the like) were very close to those of the Coal Board.

The National Union of Mineworkers had seen a fifty-year-old objective achieved in 1946 and was determined to do all that it could to ensure that nationalisation worked and was seen to work. The national officials supported Coal Board policy on a wide front and must take a considerable amount of the credit for the degree of success that was achieved by the NCB during its first few years of operation.

The reaction of individual miners, however, was by no means as clear. They too had supported the idea of nationalisation for some time but they seemed to have different notions about its likely consquences. There is little doubt that many of them were disappointed by the absence of widespread, immediate and visible changes in the structure and the management of the industry, particularly at pit level.

The miners, nevertheless, did gain considerable benefits in the 1946-56 period. Wages, in particular, increased, both absolutely and relatively far faster than they had done in the past. Whether this was due to nationalisation or the improvement in the market position of the industry, though, it is difficult to determine. Certainly a number of contemporary commentators argued that the latter factor had as much, if not more, influence than the former.

The key to the whole of the post-war period, in fact, was increased demand and pressure for expansion. The position of the industry had changed dramatically. Miners, the NUM, the Coal Board and the government all believed that to increase the level of coal production was of paramount importance. It was almost as if the cost of coal did not matter; only the quantity mattered. Few observers were willing to say that such a policy was wrong.

NOTES

1. Coal Nationalisation Act, 1946, Clause 1 (1)
2. 8 February 1947, p.252.
3. National Coal Board, *Annual Report for 1948,* NCB, London, 1949, p.1.
4. National Coal Board, *Plan for Coal,* NCB, London, 1950.
5. Ibid., p.20.
6. Ibid., p.20.
7. National Coal Board, *Investing in Coal,* NCB, London, 1956.
8. Apparently counter-productive and withdrawn in 1950 on the recommendation of the NCB.
9. National Coal Board, *Annual Report for 1947,* NCB, London, 1948.
10. National Coal Board, *Annual Report for 1950,* NCB, London, 1951.
11. Political and Economic Planning, *The British Fuel and Power Industries,* PEP, London, 1947.
12. Ibid., p.80.
13. K.J.W. Alexander, "Wages in Coal Mining Since Nationalisation", *Oxford Economic Papers,* Vol.8, No.2,1956, pp.164-81.
14. Ibid., p.177.
15. G.B. Baldwin, *Beyond Nationalisation – The Labour Problems of British Coal,* Harvard University Press, 1955.
16. Ibid., p.224.
17. F.D.K. Liddell, "Attendance in the Coal Mining Industry", *British Journal of Sociology,* Vol.5, 1954, pp.78-86.
18. National Coal Board, *Annual Report for 1947,* NCB, London, 1948.
19. W.W. Haynes, op. cit., p.140.
20. See a series of articles published in *The Times,* 22, 23, 24 November 1948. Reid made this particular claim in the article on 22 November 1948, p.5.
21. See *Times,* 23 November 1948, p.5.
22. See *The Times,* 24 November 1948, p.5.
23. See series of articles in *The Times,* 22, 23, 24 November 1948.
24. *The Times,* 23 November 1948, p.5.
25. Acton Society, *Nationalised Industry 7: The Extent of Centralisation Part II,* Acton Society, London, 1951, p.33.
26. D. Burn, "The National Coal Board", *Lloyds Bank Review,* Vol.19, 1951 pp.33-49.
27. Acton Society, *Nationalised Industry 7 : The Extent of Centralisation Part II,* Acton Society, London, 1951, p.35.
28. W.W. Haynes, op. cit., p.326.
29. Reported in *The Economist,* 3 January 1948, p.27.
30. Coal Nationalisation Act, 1946, Clause 3 (1).
31. Ibid., Clause 3 (4)
32. Ministry of Fuel and Power, *Report of the Committee on National Policy for the Use of Fuel and Power Resources* Chairman Ridley), Cmd, 8647, HMSO, London, 1952.
33. Ibid., p.68.
34. *Economic Survey for 1948,* Cmd, 7344, HMSO, London, 1948.
35. Ibid., p.3.
36. Coal Nationalisation Act, 1946, Clause 1 (4) (c).
37. Ibid., Clause 46 1 (b).
38. This was another aspect of Coal Board policy criticised by C. Reid.
39. W.W. Haynes, op.cit., p.154.
40. Ibid., p.158.
41. A Moffat, *My Life with the Miners,* Lawrence and Wishart, London, 1965, p.86.

98

42.   M. Cole, *Miners and the Board,* Fabian Society, London, 1949.
43.   Acton Society, *Nationalised Industry II : The Workers' Point of View,* Acton Society, London, 1952.
44.   Ibid., p.5.
45.   There were, of course, a number of other factors which influenced the strike record during this period. Baldwin, for example, argued that many strikes were the result of the use of the piecework method of payment.

> The conclusion is clear: the great majority of unofficial strikes begin with the coal-face grades, particularly the "direct labour" groups (colliers and fillers) although stoppages among rippers, packers, wastemen, pan-turners (conveyor shifters) are not uncommon. It is the highest wage groups (those paid under incentive contracts) which produce most stoppages and the lowest wage groups (those paid by time) which provide the fewest; it would appear the prevailing method of incentive payment underlies the great majority of stoppages. (Baldwin, op.cit., pp.86-7)

See also, W.H. Scott, *et al, Coal and Conflict,* Liverpool University Press, Liverpool, 1963 ; C. Slaughter, "The Strike of the Yorkshire Mineworkers in May, 1955", *Sociological Review,* Vol. 6, 1958, pp.241-59 : E.L. Trist and K.W. Bamforth, "Some Social and Psychological Consequences of the Longwall Method of Coal-getting", *Human Relations,* Vol. 4, 1951, pp.3.-38; and B.J. McCormick, "Strikes in the Yorkshire Coalfield 1947-1963", *Economic Studies,* Vol. 4, 1969, pp.171-97.

5.   A SUDDEN REVERSAL

Between 1946 and 1956 inland coal consumption had increased by an average of 3·2 million tons a year. Apart from one short period during the industrial recession of 1951-2 there had, for the whole of the 1946 to 1956 period, been a shortage of coal. In 1957, however, the situation changed and there was a substantial fall in home consumption; at 212·9 million tons the 1957 figure was 4·7 million tons less than it had been in 1956.

The NCB in its annual report for 1957 set out some of the reasons for the sudden reversal in the demand for coal.

> The average temperature for the year was 2·1 °F higher than in 1956 and this is estimated to have accounted for at least three million tons of the reduction in coal consumption. The check to industrial production in the early part of the year, when coal requirements are usually high, also contributed to the fall. Over the year as a whole, industrial production was slightly higher than in 1956 but the increase was mainly the result of more production in the summer when space heating requirements are at their lowest, and of greater activity in industries which are not large consumers of coal.
>
> The more efficient use of coal, and the increased use of oil also affected the consumption of coal. Up to the end of May, oil supplies were restricted, but from then on the use of oil increased at about the same rate as before the Suez crisis. During the year, oil equivalent to 16·5 million tons of coal was used in markets where it was an alternative to coal . . .
>
> Sales of gas were three per cent lower than in 1956. For this reason, and because of greater efficiency in gas production, coal consumption at gas-works fell by 1·4 million tons. With the introduction of further electrification schemes and more diesel locomotives, the railways used 700,000 tons less coal. Collieries used 700,000 tons less as the result of increased electrification, the installation of more efficient boiler plant and the more efficient operation of existing plant.[1]

All the signs, however, were not quite so bad. Consumption by the electricity industry and coke ovens was up by 0·8 million tons (2 per cent) and 1·4 million tons (4·8 per cent) respectively and there was still a shortage of large coal, supplies of which were imported from abroad.

100

As a result the Coal Board was confident that the setback in 1957 was only temporary. Thus it said,

The fall in total consumption of coal in 1957 does not of course mean that demand has reached its peak.[2]

In fact, the fall in demand in 1957 was to prove to be much more than a temporary setback. In 1957 total consumption had fallen by over 6 million tons (4·7 million tons inland, 7·8 million tons export); in 1958 total consumption fell by a further 13 million tons (10 million tons inland, 3 million tons export), and in 1959 total consumption fell again by over 13 million tons (almost all the fall was in inland demand) to little over 193 million tons. Between 1956 and 1959, then, total coal consumption fell by about 33 million tons or about 15 per cent. Part of the decline was accounted for by the contraction of the export market (exports fell by about 5 million tons) but the decline in inland consumption (over 26 million tons) was far more important.

Between 1960 and 1964 (see Table 14) the coal market steadied a little. Consumption did not increase but neither did it fall appreciably. In 1964 total consumption, at 193·2 million tons, was only half a million tons less than it had been in 1959, and inland consumption at 187·2 million tons was only 2 million tons less than the comparable 1959 figure. After 1964, however, the position began to deteriorate once more. Consumption, both inland and total, fell in every year except 1968, and between 1964 and 1970 both total and inland consumption fell by about 35 million tons. By 1970 inland consumption at 154·4 million tons was 63 million tons or 29 per cent less than it had been in 1956, and total consumption at 157·6 million tons was about 70 million tons or about 31 per cent less than it had been in 1957.

Table 15 shows that inland consumption of coal declined fairly generally across the board. Consumption over the 1956 to 1970 period, for example, declined from 12·1 to 0·1 million tons on the railways, from 5·8 to 0·8 million tons in the iron and steel industry and from 30·6 to 15·2 million tons for house coal. The only major consumer which did not follow this trend was electricity supply, whose consumption increased from 45·6 to 76·0 million tons.

The prime reason for the decline in the demand for coal was the substitution of alternative fuels. Table 16 shows the consumption of major fuels in Britain between 1960 and 1970. It can be seen from this table that while the consumption of coal fell over the 1960 to 1970 period, the consumption of other fuels, in particular oil and natural gas, increased markedly. Thus, whereas in 1960 oil accounted for only 25 per cent of total consumption in the UK, in 1970 the comparable figure was 47 per cent. Natural gas did not appear on the scene in any appreciable quantities until almost the end of the decade but by 1970 it was providing about 13 per cent as much energy as coal (direct use).

General economic considerations were undoubtedly the most important reason for the switch from coal. Many consumers simply found other fuels, in particular oil, cheaper. The price of oil, for example, fell from £8·06 per ton in 1957 to £7·28 per ton in 1966, while the average price of coal rose slowly but steadily over the same period. There was, however, one other factor which also had an important influence on the move away from coal; the Clean Air Act of 1956. Under the terms of the Act a local authority could apply to the Government to have part or the whole of its area declared a 'smokeless zone'. If such an order were made, and many were made, particularly in the 1960s, then both industrial and domestic users were prohibited from using fuel that produced 'dark smoke'. The Act was strengthened in 1968 with the passing of a second Clean Air Act which gave the Minister of Housing and Local Government power to direct local authorities to submit plans for smoke control. The Acts had an important impact on coal consumption for two reasons. First, they forced consumers to examine existing sources of power, and many concluded that they would be better off changing from coal. To a certain extent the coal-mining industry was unlucky, because many of the decisions to change from coal were based on a mistaken impression about the problems of converting existing and installing new coal burning appliances which would not produce smoke in contravention of the 1956 and 1968 Acts. Second, the Acts encouraged a heavy demand for smokeless fuel. It was not until the 1960s that the National Coal Board was able to produce smokeless fuel in substantial quantities and even then supply fell well short of demand. The absence of easily obtainable smokeless fuel derived from coal itself persuaded many consumers to switch to oil.

It also needs to be noted that some of the decline in coal consumption was not the result of a switch to alternative fuels but was the result of greater efficiency in coal burning. The steel industry used less coke (in 1959 16·7 cwt. of coke was used per ton of iron but by 1970 only 13 cwt.) and the electricity industry burned less coal (better plant and metallurgical advances allowing higher pressures and temperatures, allied to the greater size of stations and generating sets, enabled the thermal efficiency of steam stations to be raised from 25·2 per cent in 1957 to 28·4 per cent in 1970 and the saving of about 10 million tons of coal equivalent) for this reason.[3]

It is important to stress, though, that the problems facing the coal mining industry were by no means peculiar to Britain. The fall in demand for coal was world-wide and all other major producers, except for East Germany, Poland and the USA (which had considerably better

102

coal reserves), suffered just as badly as Britain. In a review of major
coal producers in 1967 *The Economist* said:

> Almost everywhere one looks coal is up against it. This week in
> Bonn, the West German government plucked up its courage and
> appointed a Dr. Beeching for the pits. It's about time it did some-
> thing. The industry is already being subsidised to the tune of £90
> million a year. ...In Belgium the Borinage has long been a profoundly
> depressed area. Belgian coal production has come down by over a
> third in ten years. . . The Comecon countries are making a
> thoroughly agonising reappraisal, to recoin a phrase, of their coal
> problems. In Czechoslovakia economists have worked out that a
> loss of rather over 30 shillings is being made on every ton produced.
> Hungary is in an even worse plight. One calculation has it that if
> Kadar kept paying the miners their wages, but closed the pits and
> imported all coal from Poland, he would save money . . . In short in
> only a very few countries is the coal industry at the present on the
> up. America, Poland, East Germany; that is almost the end of the
> list. And even in this handful current optimism (or its communist
> counterpart, planned expansion) is threatened. Foreign experts are
> amazed by East Germany's decision to push up production. In
> America the challenge from nuclear power is growing. Only in
> Poland, in fact, do the economics look right for continued
> economic growth.[4]

NCB's Reaction

The Coal Board's initial reaction to the decline in the demand for coal
was to treat it merely as a passing phase. Reference has already been
made to how the Coal Board argued that in 1957 the demand for coal
had not reached its peak. In 1958, when the decline in consumption
continued, the Board recognised that the 'slump' might last longer then
they had originally anticipated and as a result took action to restrain
production. Recruitment was restricted (at the end of 1958 total
manpower in the industry was about 23,000 less than it had been at the
beginning of the year) and Saturday working, introduced during the
coal shortage of the late 1940s, was suspended. Nevertheless the Board's
attitude was that the decline in consumption would eventually be
reversed and that it should therefore be careful not to take any
measures which would impede the industry's ability to meet the future
increases in coal consumption. Thus, the chairman of the NCB said:

> Nothing in the present situation justifies far-reaching changes
> in the basis of our long-term production and investment
> policies.[5]

The industry, he said, rather must 'plan for a return to conditions of expansion'.[6]

Despite the Board's efforts to restrain production in 1958, production continued to exceed consumption and by the end of 1958 the Board held stocks of 20 million tons of coal, compared to 8·6 million tons at the end of 1957. The level of stocks caused some concern and a considerable amount of adverse reaction from politicians and the news media. The Board, however, maintained that there was no reason to feel anxious about the level of stocks. To the contrary, a high level of stocks was a necessary precaution against the inevitable increase in the demand for coal in the future.

> Coal stocks (are) a valuable asset – ready to meet the first impact of increased demand for coal when industrial activity improves) . . . It would be a mistake, the Board believed, to turn mining's long-term plans upside down because of the present weakness of the markets.
>
> 'So far as our troubles are due to the recession we must budget for recovery.' No responsible person was in any doubt that sooner or later more coal would be needed than was being produced today.[7]

The continued decline in consumption in 1959, however, clearly served to change the Board's outlook. Production outstripped consumption to such an extent that by the end of 1959 stocks held by the Board had risen to 36 million tons, approximately 20 per cent of annual output. The Board's public statements, which had only the previous year enthused about the inevitable reverse of the current trend towards a decline in the level of consumption, showed how bad the position had become. The Board's chairman, Sir James Bowman, told newsmen in 1959: 'We are facing a struggle for the life of coal in Britain.'[8] The stability of the markets in the 1960s, though, once again raised the industry's hopes. It was clear that the industry was not going to regain all of the markets it had lost in the past few years yet there were hopes that a further decline might be averted. Lord Robens, the Board's new chairman, set the industry a target of 200 million tons of coal a year and this target was approached each year between 1960 and 1964. In 1962 the Board made its first profit for six years and the following year Robens announced that the industry was 'on the verge of a new breakthrough in cheap energy for this country.'[9] This prediction was based on the belief that if coal consumption could be raised above the 200 million tons a year mark then the cost of the extra production would be no more than £2 a ton (because of the industry's high fixed costs).

104

Initially the Board's targets and predictions were given government support. The reduction in coal consumption in 1965, however, led to a change of heart. In 1965 the government published a White Paper on Fuel Policy[10] in which it estimated the likely demand for coal in 1970 at no more than 180 million tons.

The size of the market for coal in any particular year depends on the growth of industrial activity, and on the weather and other unpredictable factors. The estimates in Section IV suggest, however, that the market for coal in 1970 may not exceed 170 to 180 million tons.[11]

This estimate was confirmed by the National Plan.[12]

Robens and the Coal Board, however, refused to be diverted from their target either by the reduction in consumption or the government's forecasts. They argued that a target of 200 million tons a year was not only realistic but also essential. Three points were stressed in support of this argument. First, it was important for Britain to retain a strong coal industry; a reliance on oil would mean that she would be economically vulnerable to action by oil producers. Thus,

fuel policy which places relaince on oil will virtually place its economic future in the hands of the Middle East and, to a lesser extent, the Soviet Bloc.[13]

Second, if less than 200 million tons of coal were produced it would, because of the high level of fixed costs, significantly increase the cost of production per ton of coal.

The coal industry pays over £43 million a year in standing interest charges before it can even begin to talk about net profits. We also carry £185 million a year of other fixed overheads and reckon to carry them comfortably on 200 million tons a year. If tonnage drops below that figure, our costs a ton would inevitably rise. On an output of 200 million tons, standing overhead charges are roughly 22s.6d. a ton; on 180 million they rise to 25s.; and on 150 million tons they would rise to 30s. The increasing productivity in the industry could not absorb soaring overheads a ton which would inevitably result from any abrupt and ill-considered curtailment of output.

But this is only half the story. The greatest part of our labour — all workers, in fact, except faceworkers — constitute a cost which does not vary with output. If the industry should lose 10 or 20 million tons of its market, and therefore 5 to 10 per cent of its proceeds, it would not reduce its costs in anything like the same proportion.[14]

Third, a reduction of production below 200 million tons a year would mean widespread redundancies. Redundancies would not only cause human problems but they would also mean that manpower would be lost to the industry for ever.

The retention of the 200 million tons a year target, however, became less realistic as the years progressed. Production in 1966 fell to 174 million tons and in 1967 to 172 million tons. By 1967, then, it had become clear that the Coal Board figure of 200 million tons a year was more of a rallying call than anything else.

In 1967, however, the Board's hopes and prestige suffered yet another blow. The government, in that year, published another White Paper on fuel policy[15] and it predicted a further decline in coal consumption in the future.

> The Government have concluded from their analysis of coal's position and prospects in relation to competing fuels that, on any tenable view of the long-term pattern of energy supplies and costs, the demand for coal will continue to decline. This is not the result of Government policy; it reflects a continuing trend in consumer preference.[16]

The estimates for coal consumption were 149 million tons in 1970 and 120 million tons in 1975.

The publication of the White Paper finally ended any hopes that the Board might still entertain of achieving their target of 200 million tons a year. Nevertheless, the Board continued to be more optimistic than the government about the future of coal. In 1970, for example, they told the Prices and Incomes Board[17] that although the government's projection was for the consumption of 120 million tons of coal in 1975 they felt that a figure of 150 million tons a year might be realistic.

NCB Plans Revised

In 1950 the NCB had published its long term plans for the reconstruction of the industry. Although the plans had been revised in 1956 to take account of increased costs the massive programme of investment had remained intact. The aim of the 1956 plan had been to produce 240 million tons of coal a year. It was clear in 1959 that despite the Board's optimism about the long-term future such a target had become unrealistic. The Board, therefore, set about a further revision of its plans.

In October 1959 the Board published its 'Revised Plan for Coal'.[18]
This revision took account of the reduced demand for coal and scaled
down the Board's investment accordingly. Introducing the revised plan
to the industry the Board's chairman said:

> When the (original) plan was first drawn up in 1950, it was never
> intended that it should be a blue-print for the next fifteen years. It
> was a guide to the reconstruction of the industry. In 1956, it seemed
> that the demand for coal was likely to be greater than could be met
> from deep-mined output and in our first revision − 'Investing in
> Coal' − we thought we should need 10 million tons a year of open-
> cast, in addition to 230 million tons from the deep mines, to meet
> demand in 1965.
>
> Now in 1959, the future looks different, and we estimate that in
> 1965 demand will be between 200 and 215 million tons − not much
> more than it is now.[19]

Output was to be concentrated almost entirely on deep mines, open
cast production contributing only about 2 million tons a year by 1965.
Production of deep-mined coal would be expanded in certain areas,
notably the North Eastern, East Midlands, West Midlands and South
Western Divisions, but reduced in others, especially the Northern
(northumberland and Cumberland), Durham and North Western
Divisions. In all this would mean the closure of 155 to 190 pits and the
merger of 50 more. The Board intended to continue with its plans for
the reconstruction of the industry, although on a reduced scale;
expenditure would be about £200 million less (at 1959 prices) than
estimated in 1956. Despite the chairman's defiant remark that

> far from winding up the coal industry, we are shaping it to play a
> part in the future of our country just as important as that played
> by King Coal in Britain's past[20]

the Board made it clear that even the reduced level of production
envisaged in the report was dependent on the maintenance of coal's
competitive position *vis-a-vis* other sources of fuel. Thus,

> we shall only be able to sell as much as this if, by efficient and
> economic production, we can keep the price of coal steady.[21]

Pit Closures

Both the original 1950 Plan and the 1956 revision had envisaged the
closure of a number of pits. They both also expected, however, that
such closures would be more than balanced out by the extension of

working at other pits and the opening of new mines. The 1959 revision changed all this. Production was not to be expanded and although new developments were planned they were not expected to be able to absorb all of the manpower displaced by pit closures. The 1959 revision envisaged a contracting industry, at least in terms of manpower and units of production, if not in terms of total output.

The contraction that took place in the early 1960s was, in fact, greater than that envisaged in 1959. Between 1957 and 1965 about 300 collieries were closed and another 30 merged. Over the same period manpower in the industry fell from 703,800 to 465,600 (see Table 17). The worst-hit regions were Scotland (production fell from nearly 21 million to about 15 million tons and manpower from 85,600 to 48,100), Northumberland and Durham (production fell from over 37 million tons to 30 million tons and manpower from 143,000 to 91,000) and the South West, essentially Wales (production fell from over 23 million tons to about 17 million tons and manpower from 104,600 to 66,700).

The closures of the early 1960s, however, were only a beginning. Because of the severe reduction in coal consumption after 1964 the NCB was forced to review and accelerate its programme of pit closures. The new programme published in November 1965 (published on the government's insistence) classified collieries into three categories. Category A pits were long-life profitable pits, of which there were 283. Category B pits were those which were not profitable but which might become so. The 100 pits in this category included a number of new pits which would become profitable when they were fully operational (it takes at least five years for a new pit to become profitable since five or six faces need to be opened before the overheads can be covered) and others which might become profitable with mechanisation. Category C pits, of which there were 150, were those where coal reserves were nearly exhausted or which were never likely to make a profit. The plan was to concentrate production by 1971 in about 310-320 long-life profitable pits which were easy to mechanise and could produce marketable coal. This would necessarily mean the closure of another 200 pits.

The programme of closures moved quickly. Between 1965 and 1969 over 200 pits were closed (see Table 18). By 1969 the number of small pits (producing under 250,000 tons a year) had been reduced drastically; in 1969 only 96 were in production compared to 256 in 1965. Output was clearly being concentrated in the larger pits; in 1969 over 21 per cent of output was produced in pits with an annual output of 1 million tons or over (compared to 10·5 per cent in 1965) and 65 per cent of output was produced in pits with an annual output of 500,000 tons and over (compared to 48 per cent in 1965). By 1969-70 coal production was down to 140 million tons and the size of the industry's labour force had been reduced to 305,100 (see Table 17). Once again certain areas suffered most; generally again, Scotland, Northumberland and Durham, and Wales. Production, for example, in 1969-70 was 11·3 million tons in Scotland (compared to 15 million tons in 1965), 20·5 million tons in Northumberland and Durham (compared to 30 million tons in 1965) and 12·8 million tons in Wales (compared to 17 million tons in 1965). Manpower in these regions showed a similar fall, to 30,600 in Scotland (from 48,100 in 1965), to 50,600 in Northumberland and Durham (from 91,000 in 1965) and to 42,600 in Wales (from 66,700 in 1965). Certain areas benefited from the Coal Board's policy of concentrating production in long-life pits, especially Yorkshire and the Midlands. Even the expanding, profitable areas, however, suffered reductions in the level of manpower, for with increased mechanisation more coal could be produced with fewer workers. The labour force in Yorkshire, for example, fell by more than 20 per cent over the 1966 to 1970 period.

One of the consequences of the reduction in the size of the industry and the concentration of production in larger units was the need to change the industry's organisational structure. The changes that took place were the first important moves in this area since the publication of the Fleck Report in 1955. In 1964 the Board set up a committee, with the Board member for Production as chairman, to investigate its own organisation. The committee reported back early in 1965. Later in the year the Board announced significant changes based on the report's recommendations.

The old Divisions and Areas were to be abolished and in future the industry was to be organised on the basis of seventeen new large areas. It was estimated that on average each new large Area would employ 20,000 men and produce 10 million tons of coal a year. The Areas were to be placed under the control of a single person, the Area Director, who was to be assisted by two deputies; a Deputy Director (Operations) and a Deputy Director (Administration). Minor alterations were also to be made to administration at colliery level in an attempt to strengthen administration and personnel work. The Board estimated that the reorganisation would lead to a reduction of 13,000 to 14,000 in management and clerical personnel by 1970, which would itself help to save £12 million to £15 million a year.

Mechanisation

The programme of pit closures was one essential element of the Board's revised plan. The other essential element was investment to increase mechanisation in the industry. Mechanisation itself involves a number of elements. The first moves to mechanise coal mining came with the use of mechanical cutters. These took a great deal of the manual work out of coal winning but they by no means eliminated it altogether. It was still necessary to blast the coal after cutting and then shovel it on to conveyors. The next stage in the mechanisation of coal mining came with 'power loading'. This permitted coal not only to be cut mechanically but also to be loaded and prepared for transport mechanically. The development of power loading on a large scale, however, took some time. Initially problems were encountered because of the great variety of conditions under which coal was mined. Variations in natural conditions – thickness and gradient of seam, faulting, the nature of the surrounding rock and so on – made it virtually impossible to design a machine that would be universally satisfactory. A breakthrough was achieved in 1952 with the invention of the Anderton Shearer Loader. This machine was simple in design but very effective in a wide variety of conditions. A further development came with the extension from power loading to powered support systems. Powered supports eliminated the time that was wasted while props were put up to enable other machinery to be brought in to start work. An indication of the importance of this development can be seen by the fact that when working with normal supports the average output per manshift using Anderton Shearer Power Loaders in 1961 was 6·58 tons; the comparable figure for the same machine when used with powered supports was 8·89 tons. Since the development of powered supports a whole host of further ancillary help has been introduced, such as improved transport and communications systems, and a start has been made with the development of remote control mining.

Table 19 gives some idea of the extension of the use of mechanisation in the coal-mining industry during the 1960s. Whereas in 1961 only 38 per cent of longwall faces (about 90 per cent of all faces are now longwall) were mechanised by 1969 the total was approaching 90 per cent. If one looks merely at the extension of the use of power-loading equipment and powered supports then the picture is even more dramatic. Prior to 1955 neither were used to any great extent (power loading, for example, only accounted for 11 per cent of total output while powered supports were not used at all) but by 1969 92 per cent of output was power loaded and 74 per cent of all faces had powered supports. Recently moves have been made to use ROLF (the remotely-operated longwall face system). This system uses the power loaders and supports previously mentioned, but is, in addition, controlled away from the face itself. It means, in fact, that mining can be carried on without employing men to work at the coal-face. This system has encountered considerable problems but has been tried out at the Bevercotes colliery in Nottinghamshire.

The Coal Board's programme of pit closures and its ettempts to increase mechanisation to a large extent paid off in terms of productivity; output per man shift increased significantly in the 1960s. Table 20 shows the increase in output per man shift over the period under consideration. It is clear that the rise in output had its peaks and troughs (the best time for output was probably in the early 1960s and between 1967 and 1969 — some commentators argued that trend towards increasing productivity of the 1960s was'petering' out by the turn of the decade), and it should also be remembered that the increase in productivity varied markedly from region to region (the increase was highest in the South Midlands area). Nevertheless, the general increase over the period, throughout the industry, was impressive, certainly by past standards when productivity had increased very little.

Marketing of Coal

The Coal Board hoped that by increasing productivity it could improve the industry's competitive position. Only if the industry improved its competitive position, they argued, could it hope to increase coal sales. Although it is clear, then, that increasing productivity was seen by the Board to be of prime importance, the Board also attempted to increase coal sales through measures taken in two other directions. Obviously the price of coal had an important bearing on its attractiveness to customers; the industry could not, however, afford to ignore the effect of marketing and distribution.

Ever since the Second World War there had been restrictions on the sale of coal. Industrial consumers had since the end of the war received allocations from the Ministry of Fuel and Power, while domestic consumers had been restricted both in the amount of coal they could purchase and in the merchant from who they could purchase the coal. In 1958 the control of supplies came to an end and for the first time the Board had to engage in selling coal rather than merely sharing it out.

The Board's first efforts were directed towards improving the 'image' of coal. In 1959 the Coal Board's Director-General of Marketing told a conference of coal traders:

> Coal has a great deal to commend it. We know it, but do the the public know it? We haven't bothered telling them for a great many years.[22]

The coal industry had got make an attempt, he said, to offset the feeling that oil was the 'fashionable' fuel. In 1961 Lord Robens, Coal Board Chairman, listed some of the measures taken in this direction:

Since the start of April (1961), anybody who still clung to the belief that coal was a back number has had a rude awakening. Throughout Britain, as a result of a massive publicity campaign nobody is in any doubt that coal means business . . . The facts are there. Our job is to make them known, using every medium that is open to us – advertising, TV, films, exhibitions, press conferences and so on. Headquarters and the various Sales Regions are combining to make this an all-out effort. Our advertising is the spearhead of the drive, but many supporting activities are taking place throughout the regions.

By the end of May, Housewarming Plan advertisements will have appeared in national newspapers with a total of 40,000,000 readers, as well as in 400 local papers. Appearances of our industrial advertisements, based on the theme 'Progressive Industry is Going Forward on Coal', have been stepped up. Throughout the whole campaign special emphasis is to be given to 'Sunbrite', the NCB hard coke.

A major event is the five-week central heating drive which we are undertaking jointly with the Coal Utilisation Council. Our campaign includes 26 exhibitions and more than 50 show houses featuring solid fuel heating up and down the country. To ensure maximum publicity for these, as many as possible are being opened by well-known people.

The Regions are all making special efforts to back up the campaign. There will be a solid fuel brains trust in Newcastle with a stage or television celebrity on the panel. In Sheffield a central heating competition is being run by the *Sheffield Star*. The Spalding Tulip Parade will include a Coal Board float. A half-hour television feature has gone out from Kent.

Buses, NCB lorries and London's Underground trains all carry our posters at dominant sites throughout the country – 600 in Yorkshire alone. Their theme, and that of the whole campaign, is embodied in this slogan:

"Solid Fuel – More Heat At Less Cost – And its British!".[23]

It was clear, however, that if the Board was to be successful in its efforts to persuade householders and firms to buy more coal then it had to do more than merely improve the industry's image'. The switch to oil was not only the result of the industry's bad image but also the result of utilisation problems. Many consumers, for example, were unaware of how their appliances could be converted to conform with the smoke control requirements. To a certain extent problems such as these could be overcome by providing more information. Thus the

Board gave considerable support to two bodies operating in this area, the National Industrial Fuel Efficiency Service (which gave advice on the use of coal in industry) and the Coal Utilisation Council (the main advisory service for householders). For example, the Coal Utilisation Council's promotion budget was, with Coal Board support, increased in 1959 from £100,000 to £300,000 a year. This problem could not, however, be completely solved in this way, for it was not only more information that was needed but also more and better appliances. The Board's entry into this field came rather late, but it was probably more imaginative. In the mid-1960s the Board joined with a number of private firms to promote the development and production of better coal-burning appliances. In 1964, for example, the Board agreed with J.H. Sankey and Son Ltd. to form a new company to supply solid fuel heating equipment and to act as heating consultants and builders' merchants while in 1966 the Board acquired a 50 per cent shareholding in Bradley's (Concrete) Ltd, a manufacturer of coal-burning appliances, to promote the use of colliery shales.

Distribution of Coal

The Board in the early 1960s controlled all significant coal production in Britain. It did not, however, have such comprehensive control over the distribution and sale of coal after it left the pithead. Yet it was clear that the transfer of coal from the pithead to the consumer was of vital importance; the efficiency of this operation could determine both the price of coal and the 'image' the consumer obtained of the industry.

The Board's first moves were to deal with the distribution of coal from the pithead to the individual retailer. Although road, sea and canal were used, rail was by far the most important source of transport for this part of the journey (in 1965, for example, 109 million tons or 62 per cent of Britain's coal output was carried by rail, compared to 21 per cent by road, 12 per cent by sea and 1 per cent by canal). The system of distribution by rail was far from satisfactory or efficient. A Fabian Society report published in 1956[24] described how the operation was typically conducted.

In a typical case, coal wagons arrive at the railway stations of big towns in a goods train made up of wagons containing coal, potatoes, steel, cattle, scrap iron, small parcels, furniture, etc. — what railwaymen call 'rough traffic' — or sometimes in complete train-loads carrying some 500 tons of coal. On arrival at the yard or depot the wagons are sorted out and dispatched down the numerous special sidings to individual merchants who have rented sidings from the railways. The unloading is nearly always done by primitive hand methods. The coal is loaded into hundredweight

113

bags by two men, standing in the wagon (as soon as space has been cleared) one of whom fills the bag with a shovel while the other holds it open, and when it is full puts it on the scales and weighs it. The bags are then either put on the ground to await the lorry, or they are carried or wheeled on a trolley to the lorry which is backed up to the wagon.[25]

This quotation highlights two main problems. First, there were problems of transport itself; second, there were problems associated with the coal depots. Although about half of the railway's fleet of freight wagons was committed to the coal trade most of it was organised on a fairly haphazard basis. In the majority of cases coal wagons were normally only part of the cargo being moved; as the above quotation pointed out, some were moved as complete train loads, but most were part of 'rough traffic'. This meant that a great deal of time was lost in marshalling, turn-round and the like. Dr. Beeching, chairman of the British Transport Commission, estimated in 1962[26] that of the eleven days a wagon was in use to transport one load of coal from the pithead to the depot, five or six of these days were taken up while the wagon stood empty at the pits or with consumers, four to five days were taken up in marshalling yards, while only half a day was required for actual movement. In an attempt to improve this position, therefore, the Coal Board and British Rail began separating coal from other traffic and introducing what became known as the 'coal merry-go-round'. This latter idea was quite simply a system by which coal was supplied by one colliery or one group of collieries to one customer or depot on a regular basis. Coal would be carried as a single, instead of as before a mixed, load and if possible the same wagons would be used constantly on the one run. Plans were made so that by 1971 15 million tons of coal would be moved in such a fashion.

The problem of depots had been looked at by the Robson Committee in its report[27] published in 1958. The committee had argued that a great many of the industry's distribution problems arose from the use of too many, poorly organised, unmechanised depots. The solution, which the Board started to apply in the 1960s, was to concentrate coal deliveries to a few, modern, mechanised depots.

The first such depot was opened in late 1963 at West Drayton. The depot, which was jointly sponsored by the NCB, British Rail and Stephenson Clarke (coal wholesalers and shippers), cost nearly £400,000 to build. It was situated thirteen miles due west of Paddington Station and designed to deal with most of the traffic from South Wales and the Midlands to the London area. With its up-to-date equipment it was able to handle 200,000 tons of coal a year — 130,000 tons of house coal from the Midlands and 70,000 tons of anthracite from South Wales.

114

The depot at West Drayton was, in fact, only the first of many. By 1966 half of all domestic solid fuel was passing through 'concentration depots' and by 1970 54 fully mechanised and 411 partially mechanised depots were in operation.

The other weak link in the distribution chain concerned the retail distributor. This part of the trade was best characterised by its enormous diversity. The trade was carried on by over 20,000 units, varying from large merchants (such as 'Charringtons' and 'Rickett Cockerell's') to numbers of small coal dealers. The NCB itself was engaged in the trade in a small way (accounting for about three to four per cent of the trade) through its retail outlets, as were the Area Gas Boards who disposed of about 13 million tons of coke and breeze a year. The major problem was that too many of the retailers were too small to stock adequate quantities and varieties of coal and were unable to offer adequate back up services.

In 1962, therefore, the NCB and the Government sponsored the Approved Coal Merchants Scheme. In order to join the scheme retailers had to agree to a 'code of practice' under which they had to declare the quality of sales, agree to give priority to urgent cases of need, and to provide the consumer with adequate information. The aim was not to restrict supplies of coal to members of the scheme; rather it was to give certain merchants a 'badge of respectability' and indicate to the consumers the retailers to which they should go for 'approved' and 'reliable' service.

The Government's Reaction

In the immediate post-Second World War years the government had pressed the Coal Board to expand output as quickly as possible. Increased coal production was essential, they had argued, for the recovery of general industrial production after the war. The downturn in demand in 1957 understandably took the government by surprise and like the Coal Board their initial reaction was to assume that it was merely a temporary recession. The Coal Board, of course, did not significantly change their advice in 1958. Nevertheless, it is surprising that the government accepted this advice for they did not have the same interest as the Board in maintaining the level of production and a number of observers were warning that a more fundamental change had taken place. The Select Committee on Nationalised Industries reporting in 1958,[28] for example, argued that plans for coal production ought to be speedily amended to take account of the new pattern of demand.

By 1959 the government, like the Coal Board, accepted that the changes of the past few years were more than a temporary phenomenon. At first they reacted by merely approving Coal Board plans for a scaling down of production and investment. After 1959, however, government

115

policy towards the coal industry moved in two different directions. First moves were made to help the industry adjust to the new situation and cushion some of the worst effects of the contraction in size. Second, the government tightened its control over the industry and encouraged the adoption of reduced output targets.

Moves to aid the industry were started by the Conservative Government in the early 1960s. Action was taken, for example, to virtually ban coal imports and in the budget of 1961 the coal-mining industry was given a measure of protection by the imposition of a 2d. a gallon tax on fuel and heating oils.

When the Labour Government came to power in 1964 the Coal Board received further help. First, in April 1965, moves were announced to increase the consumption of coal in certain other nationalised industries. The electricity industry was to give preference to coal and increase its stocks while the gas industry increased its 1965-66 estimates for coal consumption. The government also agreed to give coal a 5 per cent preference in the heating of government buildings. Second, the government relieved the industry of the oblication placed on it by the Conservative Government in 1963 to achieve a £10 million surplus to cover depreciation. Third, in November 1965 the government issued a White Paper[29] on the finances of the coal-mining industry. In the White Paper the government proposed to write off part of the industry's capital debt. The industry, the White Paper said, had become over-capitalised because of a programme of investment started in the 1960s which had been supported and encouraged by various governments.

> In the earlier 1950s there were strongly supported views — endorsed by the then government — that the market for coal might be as large as 240 million tons a year. The Board was accordingly encouraged to undertake the renovation and development of the industry to meet the expected market needs. As a result of the investment embarked upon, it now has to carry a burden of debt which hampers its efforts to hold markets and contain its costs. Some of this investment was in projects known to be risky, but it was nevertheless undertaken to provide tonnages believed to be necessary in the national interest.[30]

The government, therefore, believed that the industry should not be penalised for what to a large extent amounted to officially approved investment in the national interest. It announced its intention to write off about £400 million of the industry's capital debt. Effect was given to these intentions through the Coal Industry Act of 1965. The White Paper also proposed that the Board's borrowing limit should be raised to a maximum of £750 million and again gave effect to this through the Coal Industry Act of 1965. A fourth measure taken by the government to help the Coal Board concerned the price of coal. In July 1965 the Board proposed to increase the price of coal (with effect from 1 September 1965) to give additional revenue of £80 million in 1966-7. This would, in fact, have been the first general price increase for coal since 1960 (selective increases were made on a number of occasions leading to an average rise of 3½ per cent in the price of coal since 1960). The Board, however, at the government's request, agreed to defer the proposed price increases so that they could be examined by the Prices and Incomes Board. In the meantime, the government gave extra financial help to the industry, totalling £25 million, to compensate for loss of revenue (the Prices and Incomes Board published its report[31] in February 1966 approving the price increases and the government agreed to permit the Coal Board to raise their prices accordingly) Fifth, the government gave certain help to the industry to deal with the problems arising from the contraction in size. The Coal Board was given assistance to meet the cost of redundancy payments, resettlement expenses, and the like. The help was foreshadowed in the 1965 White Paper on Fuel Policy[32] and introduced as one of the provisions of the Coal Industry Act 1965.

In many ways, however, the industry's problems became worse rather than better over time. Coal consumption declined substantially after 1965 and the industry's finances suffered accordingly. It soon became clear that the package of measures introduced in 1965 to help the industry would be insufficient and two years later the government was forced to intervene once more.

The White Paper on fuel policy issued in 1967[33] envisaged a drastic reduction in coal consumption and production. In order to compensate and cushion the effects of this further run-down of the industry the government in the White Paper proposed to offer further help. Aid was to be given in four ways. First, the industry's borrowing powers were to be raised; in 1965 they had been raised to a maximum of £750 million, in 1967 it was proposed that they should be raised again to a maximum of £950 million. Further, the limit imposed on the Board's accumulated deficit was to be raised from £30 million to £50 million. Second, the White Paper said that the help given to the industry in 1965 to deal with the effects of pit closures should be continued and expanded in scope. Third, the electricity and gas industries were again asked to use more coal. The White Paper argued, however, that if the electricity and gas industries were asked to use more coal then they should be compensated for doing so. The cost of using more coal should fall on the taxpayer, not on the consumer of gas and electricity. It was proposed, therefore, to permit the government to contribute up to

117

£45 million for this purpose. Fourth, the White paper looked at the effect of pit closures not only on the mining community but also on the community at large. It argued that the loss of jobs in the mining industry could have an adverse effect on 'development areas' where employment opportunities were already poor. The government therefore should be permitted to give temporary financial assistance to the coal-mining industry so that the programme of pit closures could be delayed. Effect was given to these proposals through the Coal Industry Act of 1967.

Moves to alleviate the worst effects of the contraction in the size of the coal-mining industry, then, were an important part of government policy in the 1960s. It is clear, however, that the major impact of government policy was felt in another direction. At the same time as it was providing help to cushion some of the worst effects of the contraction in size the government tightened its control over the industry and ensured that the level of production was reduced even further.

The government's first important moves in this direction came through finance. The Coal Board had been instructed through the Nationalisation Act to balance its accounts, averaging out good and bad years. In general the Board had managed to do this over the first ten years of its existence. Apart from 1947, 1952 and 1955, the Board had very nearly broken even or made a profit in all other years. In 1956, in fact, the Board had managed to make a profit of £12·8 million. The downturn in the markers, allied to the heavy investment programme embarked upon by the Board, changed all of this. In 1957 the Board made a loss of £5·3 million and between 1957 and 1961 the Board's losses totalled almost £70 million.

In 1961 the government introduced its White Paper[34] on the financial obligations of nationalised industries. The White Paper dealt with nationalised industries in general although in the climate of 1961 it was clear that the coal-mining industry was one of its prime concerns. In this context it might be noted that the coal-mining industry had the largest accumulated deficit of any nationalised industry after the railways. The government, the White Paper suggested, had two major aims as far as the nationalised industries were concerned. First it aimed

to ensure that the industries (were) organised and administered efficiently and economically to carry out their responsibilities and that they (were) thus enabled to make the maximum contribution towards the economic well-being of the community as a whole.[35]

Second, it aimed to ensure that the nationalised industries despite their 'obligations of a national and non-commercial kind'[36] were not regarded 'as social services absolved from economic and commercial

118

justification'.[37] In order to achieve these aims the government announced its intention to tighten up the financial rules governing the operations of nationalised industries. First, it stated that the period over which profits and losses might be balanced out was to be limited to five years. Secondly, the industry should normally, over this five-year period, not only balance its accounts but also ensure that it had sufficient surplus to cover depreciation and provide a contribution towards investment.

In 1963 after discussions with the Minister of Power the government set the Coal Board the target of a £10 million annual surplus which would be used to provide for depreciation. One of the important effects of this target was that the Board's efforts to concentrate production had to be increased and their closure programme accelerated.

While in opposition the Labour Party frequently criticised the Conservative Government for their policy towards the coal-mining industry. They declared that when they gained power they would take steps to increase the market for coal. Thus, Wilson, Shadow Chancellor of the Exchequer said:

> Our fuel policy is this – our home-produced coal must come first. Second should come British refined oil. Imported oil must come last. Our policy for increasing industrial production will mean a growing demand for coal.[38]

In fact, although the Labour Government which was returned to power in 1964 gave considerable help to the coal-mining industry in other directions, they continued the policy of encouraging a contraction in size. The White Paper on Fuel Policy of 1965,[39] the National Plan of 1965[40] and the White Paper on Fuel Policy of 1967,[41] all envisaged a significant reduction in the demand for and consumtion of coal. Even more important, the targets set in these documents for coal production were all well below those supported by the National Coal Board and the government after 1965 was engaged in a constant battle to persuade the industry to lower its sights. The government frequently said that the figures quoted in the White Papers and the National Plan were only targets and not instructions, but as time passed it became increasingly clear that the industry would be unable to deviate greatly from the government's plans.

The Miners' Reaction

The miners clearly suffered considerably as a result of the actions taken by the NCB to reduce production in the 1960s. The impact of the NCB's actions was probably most spectacularly visible in the field of pit closures. Reference has already been made to the run-down of manpower in the industry from 658,000 in 1959 to 305,000 in 1970.

Quite considerable efforts were made both by the government and the Board to alleviate the social problems caused by this run-down. In 1965, for example, the government agreed to give up to £30 million to help displaced miners. The Coal Industry Act of 1965 listed some of the uses to which this money might be put; redundancy payments, compensation for loss of superannuation benefits, retirement payments before the normal retiring age, removal expenses, travelling expenses, housing aid, temporary supplementation of earnings and maintenance of social welfare facilities. Further aid was offered in 1967. Under the 1967 Coal Industry Act a scheme was introduced to help miners over the age of 55 who were made redundant. The scheme provided that any miner aged 55 or over, but less than 65, who had achieved a minimum period of service, would be able to claim a supplement to enable them 'to adjust to their new circumstances'. The supplement would mean, in effect, that they would receive, when taking unemployment benefit in to account, about 90 per cent of their previous take-home pay. The supplement was to be paid for a maximum period of three years, but after that time an ex-miner who was still unemployed would be able to claim a miners' pension without waiting for the normal qualifying age of 65.

The Coal Board made similar efforts to ease the situation. Wherever possible displaced miners were offered alternative employment within their own area. As the pace of closures increased, though, this became more and more difficult and many miners had to accept transfers outside their own area if they wanted to stay in the industry. If, however, they were willing to make such a move then the Board did all they could to help them. They were given the greatest possible choice of work locations (the Board used mobile vans to promote a 'Pick-Your-Pit' scheme) and help with the removal procedure.

Nevertheless, there is no doubt that despite the efforts of the government and the Coal Board, the miners suffered considerable hardship because of the programme of pit closures. The number of men actually made redundant, especially in the early 1960s, was small, but three factors made the position with regard to redundancy worse than it might appear at first sight. First, many of the men made redundant were, because of their age, those who had the least chance of obtaining alternative employment. Thus, a study of pit closures in Northumberland[42] found that 78 per cent of the men who had been made redundant were aged 56 or over. Second, the task of obtaining alternative employment was made more difficult because many miners had little experience of other work. Many miners had been employed in the coal-mining industry all of their working lives and few had extensive experience elsewhere. Thus, a study of pit closures in

120

Cumberland[43] found that only 47 per cent of the miners interviewed had any experience of work in another industry and only 21 per cent had spent more than half of their working lives in other industries.

Third, the areas that suffered worst through pit closures, Scotland, Northumberland, Durham and Wales had few alternative sources of employment to offer and, especially in the late 1960s, already a high rate of unemployment before taking the pit closures into account. The result of this combination of factors was that although only a small proportion of miners displaced because of pit closures suffered redundancy, those miners who were made redundant frequently found it extremely difficult to obtain alternative employment. This was shown in a study conducted for the Ministry of Labour and the National Coal Board of the pit closure at Ryhope, Northumberland.[44] Although only 22·5 per cent of the displaced miners were made redundant, 62 per cent of those who were made redundant were still out of work six months later.

Even the most fortunate of the displaced miners, those who were transferred to other work in the same area, did not escape unscathed. There is a considerable body of evidence to show that they suffered in a variety of ways. Thus, one study[45] showed that 82·5 per cent of the men so transferred had a longer journey to work, 52 per cent suffered a decrease in leisure time and 49 per cent lost contact with former workmates (usually not replaced by new contacts at the new workplace).

Although pit closures may provide the most dramatic evidence of the way that miners were affected by the rundown of the industry it is clear that they were also adversely affected in other ways, in particular earnings. After the Second World War, under the impact of coal and labour shortages, miners' wages rose dramatically, far faster than those of other industrial workers. After 1956, however, the position was reversed.

The miners, throughout the 1957-70 period, of course, managed to obtain fairly regular wage increases. In 1962, for example, a delegate conference of the union accepted an offer of wage increases ranging from 7s.6d. to 10s. a week; in 1964, following a ballot of members, the NUM accepted wage increases of between 3 per cent and 4 per cent (7s.6d. to 12s. a week); in 1965, again following a ballot of the membership, wage increases of between 4½ per cent and 5 per cent were accepted; and in 1966 miners obtained general weekly increases of 12s. (for underground) and 10s.6d. (for surface workers) along with the National Power Loading Agreement.

Relatively, though, despite these increases, miners' earnings fell behind those of other workers. Between 1956 and 1970 miners' earnings on average rose by about 90 per cent (see Table 12) whereas the average earnings in manufacturing industries rose, over the same period, by about 135 per cent (see Table 21).

It is, of course, difficult to say exactly what part the decline in the demand for coal, and the reduction in the level of coal production, played in the relative decline in the level of miners' earnings in the 1960s. It is clear, however, that the miners' bargaining position was considerable weakened by the decline in the industry's fortunes. They could no longer point, as they had in the late 1940s and early 1950s, to the need to attract more manpower to the industry as an argument for increasing miners' wages, and they were dealing with an employer who was having to face increasing financial difficulties, trying to sell a commodity which seemingly few people wanted and having to depend on government aid to maintain its viability.

In the immediate post-war years the coal-miners' union, and (possibly to a lesser extent) the coal-miners, had given the Board a great deal of support. Despite the change in conditions after 1956, the unions, and to a large extent the miners, continued to give the Board such support.

Coal Board Chairman Robens, for example, has told how the unions gave the Board a great deal of support in their efforts to improve the industry's productivity and increase the extent of mechanisation. Speaking at an International Management Congress at New York, he told how the questions of productivity and mechanisation had been discussed by men, unions, and management, throughout the industry, in the various consultative committees. The union's attitude had always been constructive for they had been as concerned as the Board to improve the industry's competitive position as much as possible. Robens concluded by saying:

> I believe that the working relationship of the Board and the various unions has contributed more to the rebirth of the British coal-mining industry than any other single factor.[46]

The Wilberforce Report,[47] in a similar fashion, paid tribute to the co-operation shown by the unions and the miners. They made particular reference to an issue related to productivity and mechanisation, the introduction of a new wages structure for a large section of the industry industry's work force. The first major attempt to restructure wages in the industry had been made in 1955 with the introduction of a new scheme for daywagemen.[48] It had always been envisaged that at a later date a similar scheme would be introduced for piece-workers. Action eventually had to be taken in this area, however, because of the increase in mechanisation. It was thought impossible to carry through mechanisation to the extent desired and yet retain the piecework system. The piecework system, after all, was based on the notion of rewarding men for increased effort. How could this be operated when the amount of coal produced, and to a large extent the speed of work, would be determined by the machine being used and not the physical effort of the workers? In June 1966, therefore, the National Power Loading Agreement was introduced. The agreement, which covered about 100,000 faceworkers on mechanised faces, provided for the progressive replacement of piecework by a daywage system eventually based on a standard national shift rate (initially district shift rates were used). By the end of March 1968 the agreement had been applied to 80 per cent of all mechanised faces.

One of the consequences of the agreement was that most faceworkers suffered a relative reduction in their level of earnings. Clegg, in his evidence to the Wilberforce Inquiry, described how this had happened:

> One of the consequences of the Power Loading Agreement was to hold back the pay of most face workers in relation to the rest of the industry, and to hold back the pay of some groups of face workers far more than others. It is, I believe, generally agreed by by practitioners of industrial relations that, when a new pay structure alters the relative pay of groups of workers in an industry or undertaking, there is need for a generous overall increase to be injected, so that all, or all but a few of, the workers who suffer a relative setback should, nevertheless, receive a net advance. But over the last few years a large number of miners have been asked to take a reduction relative to the general level of miners' earnings at a time when that general level has been falling behind the movement of pay in the country as a whole.[49]

The Wilberforce Committee in its report praised the miners for the way they had approached this matter. They had accepted sacrifices in the interest of greater efficiency.

> There has been quite exceptional co-operation shown by the miners in the last few years in moving from piecework schemes to day-working schemes in the interest of greater efficiency. This co-operation has been a model to industry as a whole.[50]

One of the consequences of the NUM's support for NCB policy and their refusal to oppose it on any significant scale in the late 1940s and early 1950s had been signs of frustration among rank and file members who felt that their interests had been forgotten or were being ignored. The marked increase in the number of unofficial strikes in the immediate post-Second World War period was undoubtedly partly (although of course not entirely) a consequence of this kind of feeling. It is interesting that between 1959 and 1970 the same response was not

123

evident..There were in fact significantly fewer strikes in the industry over this period.

Table 22 shows that there was a gradual reduction in the strike figures between 1959 and 1964. Thus in 1959 there were 1,307 stoppages and 363,000 man days lost through stoppages in the industry. The comparable figures for 1964 were 1,059 stoppages and 302,000 man days lost through stoppages. After 1965, though, there was an even more dramatic decline in the strike statistics. By 1968 there were only 227 stoppages in the industry and 57,000 man days lost through stoppages. Towards the end of the period there was something of a reversal of earlier trends. The number of man days lost through stoppages increased to over one million in both 1969 and 1970. This reversal, however, was in the main due to two large scale unofficial stoppages. In 1969, for example, there was a major unofficial strike over hours of work and in 1970 there was another unofficial strike over a national pay claim. Each of these stoppages accounted for about one million working days lost. Further, the number of stoppages in the industry continued to fall in both 1969 and 1970; in 1970 there were only 165 stoppages.

It would, of course, be foolish to argue that the fall in the number of stoppages, and to a lesser extent the number of man days lost through stoppages, was necessarily indicative of an increase in the extent of rank and file support for the NUM's and the NCB's actions. There are many other factors which could, and some of which undoubtedly did, affect the industry's strike record.

First, there is the simple fact of the reduction in the size of the industry's labour force. The number of men employed in the industry in 1970 was less than half of what it had been a little over a decade earlier. Table 23 shows that when the industry's strike record is weighted to take account of the decline in the size of the labour force the decline in the level of strike activity is not quite as dramatic as it might at first appear. Nevertheless, the same general trend remains even if in a less extreme form.

Second, there is the decision in 1966 to gradually replace piecework in the industry with a system of daywage payments. The way in which piecework can precipitate, if not cause, strike action has been well described elsewhere.[51] Piecework had been the reported cause of about 40 per cent of all days lost through major stoppages in the industry between 1957 and 1965.[52] The National Coal Board clearly had this in mind when they negotiated the agreement. In their annual report for 1966-7 they introduced the agreement by saying how they hoped it would lead to a reduction in the number of disputes in the industry.

The agreement will gradually replace a large number of local agreements which incorporated some element of piecework. This change to a daywage system in an industry with a long tradition of piecework represents a major advance whcih might well set a pettern for other industries. Piecework has become increasingly inappropriate as a means of payment for men on mechanised faces where productivity depends less on physical effort than on the utilisation of machines. Agreements based on piecework, moreover, needed constant revision and renegotiation and they were the source of many stoppages and disputes in the industry.[53]

Third, there is the effect of the run-down of the industry. There is little doubt that many workers were unwilling to engage in strike activity when they thought that they might lose their jobs in the near future because of the closure of their pit. One should not take this argument too far, though, for the evidence is somewhat contradictory. The most strike-prone parts of the industry remained Scotland, Yorkshire and South Wales. One of these regions, Yorkshire, remained releatively prosperous throughout this period, yet Scotland and South Wales suffered considerably from the contraction in the size of the industry. Certainly the fear of redundancy and low strike activity have not always been synonymous in the coal-mining industry and it is doubtful if the former is a sufficient explanation for the latter.

Further, it should be emphasised that even if the industry's strike record did not indicate wholesale dissatisfaction with the NUM's acceptance of NCB policy there was some evidence of dissatisfaction from other sources. In 1964, for example, a report was published, edited by the Secretary of the Derbyshire area of the NUM and the President of the South Wales area, entitled 'A Plan for the Miners'.[54] The report was extremely critical of both NCB policy and of the attitude of the NUM towards it. The report claimed that many of the Board's financial problems were more imaginary than real. The NCB, the report argued, had 'cooked their accounts' by allowing more than was necessary for depreciation. Despite the NUM's vigorous defence of the NCB's integrity and policy there can be little doubt that many miners shared the report's scepticism, even if their scepticism was not as firm as that of the authors of the report. The NUM also encountered antagonism from one particular section of their membership, the winding enginemen. An attempt was made to form a separate Winding Enginemen's Association, although it should be stressed that this particular dispute owed more to the union's structure than its policies towards the NCB.[55]

It is clear, then, that there was a certain amount of criticism from rank and file members of NCB policy and of the NUM's attitude towards it. Nevertheless, it is doubtful if this criticism was any greater during

125

the 1959-70 period than it had been earlier, and it was probably somewhat less. There is a great deal of evidence to support the argument that most of the NUM's and the miners' antagonism was directed towards the government and the political parties rather than towards the NCB. The demands made by the miners in their seven-point plan,[56] for example, nearly all called for government rather than Coal Board action.

> Financial help to the NCB towards the cost of stocking coal and of redundancy and resettlement; provision of alternative industries in mining areas; further cuts in open-cast mining; reduction in the use of oil fuel in electricity generation; increased use of coal in gasworks; re-examination of the programme of atomic energy power stations; restrictions on the dumping of fuel oil.[57]

The miners and their unions were clearly aware that if their problems were to be dealt with adequately then it was the government who had to be goaded into action. The NCB, for their part, were just as keen as the miners to take action to improve the industry's current position and its future, and spared no efforts in this direction.

Conclusion

There is little doubt that the decline in coal consumption in 1957 took most people by surprise; few had predicted such an event. As one might expect both the Coal Board and the miners were slow to react to the changes that were occurring; one can understand why both were loathe to admit that long-term changes were in train. The government, however, did not have a vested interest in a high level of coal production and they were warned by observers that the reduction in the demand for coal was likely to prove to be more than a temporary phenomenon. Yet the government was almost as slow as the miners and the Coal Board to react to changing market conditions.

From 1959 onwards the National Coal Board took a series of measures designed to deal with what they by then realised was a far more permanent reversal in the industry's fortunes. The first of such measures was the revision of the plan for the reconstruction of the industry. Although the plan was by no means abandoned output targets were scaled down and the level of investment was reduced accordingly. The moves which followed produced a much more highly mechanised and concentrated industry. Yet, despite the increase in the industry's productivity and despite the efforts made to improve the Board's marketing and distribution facilities, by 1970 the industry was producing and selling far less coal (about 30 per cent less) and providing employment for fewer men (more than 50 per cent less).

The government, for their part, at first merely encouraged the Coal Board. They supported the Board's 1959 revision of the long-term plan and the moves to reduce the size of the industry and increase mechanisation. In the early 1960s, however, the government began to intervene far more in an attempt to persuade the Board to take more drastic action. Both the Conservative and Labour Governments began to realise that the industry would not, without considerable support, be able to sell anything like the amount of coal that the Board intended it should produce. Both governments were willing to give a measure of support to the industry. Thus, for example, the Conservative Government introduced a tax on fuel oil and restricted the import of foreign coal, while the Labour Government gave considerable financial assistance to the industry as well as trying to persuade, or if necessary instruct, certain key customers (such as the electricity industry) to use more coal. Neither the Conservative nor the Labour Governments, however, were willing to take their support far enough to enable the Board to maintain production at, for example, 200 million tons a year, as it had aimed to do for much of the 1960s. Essentially both Conservative and Labour Governments were willing to help to cushion some of the worst effects of the run-down of the industry; neither were willing to help to reverse the switch from coal to oil on a more permanent basis.

The miners, at first, like most other parties, were bewildered by the dramatic change in the industry's fortunes. Later, however, they co-operated to a surprising degree in the efforts made by the National Coal Board to deal with the situation. The spirit of co-operation between the NUM and the Coal Board which had been fostered during the years immediately following the nationalisation of the industry was, by and large, maintained despite the strains imposed by the new situation. The NUM gave considerable help and support to the Board in its efforts to improve the efficiency of the industry both through the closure of pits and the increase in mechanisation.

The attitude of the rank and file members was by no means as clear-cut as that of the NUM leadership. The miners, after all, in the space of a few years had seen a change from a situation where coal and coal-miners were in demand to a situation where both were in surplus. They were stunned by the change in the industry's fortunes as they were by the programme of pit closures. Yet there is little evidence that there was any major increase in hostility on the part of the miners, either to the Board's actions or to their union's support for most of these actions.

By the end of the 1960s the position of the coal-mining industry clearly gave cause for concern. The industry and its employees had suffered a great deal throughout the 1960s and few seemed to have confidence in the industry's future. One consequence of this was that the industry was unable to attract young recruits. Allied to the run-down of manpower this had meant an ageing work force. Thus, the average age of the work force increased from 41·6 to 43·9 years between 1960 and 1970 and in 1970 only about 12 per cent of the work force was aged under 25 while nearly 40 per cent was aged 50 or over. Few industries, especially those that feel the need to increase productivity significantly in order to become competitive, can be happy about such a situation.

NOTES

1. National Coal Board *Annual Report for 1957,* NCB, London, 1958, pp.28-9.
2. Ibid., p.33.
3. See K. Allen, 'The Coal Industry', in G.L. Reid, K. Allen and D.J. Harris, *The Nationalised Fuel Industries,* Heinemann, London, 1973.
4. 27 May 1967, p.922.
5. *Coal,* July 1958, p.18.
6. *Coal,* October 1958, p.19.
7. *Coal,* November 1958, p.19.
8. *Coal,* February 1959, p.5.
9. *The Times,* 13 June 1963, p.9.
10. Ministry of Fuel and Power, *Fuel Policy,* Cmnd. 2798, HMSO, London,1965.
11. Ibid., pp.16-17.
12. Department of Economic Affairs, *The National Plan,* Cmnd. 2704, HMSO, London, 1965.
13. *Coal Quarterly,* Spring 1963, p.23.
14. *Coal Quarterly,* Spring 1963, p.23.
15. Ministry of Fuel and Power, *Fuel Policy,* Cmnd. 3438, HMSO, London, 1967.
16. Ibid., p.44.
17. National Board for Prices and Incomes, *Coal Prices,* (2nd Report), Cmnd. 4455, Report No.153, HMSO, London, 1970.
18. National Coal Board, *Revised Plan for Coal,* NCB, London, 1959.
19. *Coal,* November 1959, p.5.
20. *Coal,* November 1959, p.5.
21. *Coal,* November 1959, p.5.
22. *Coal,* November 1959, p.19.
23. *Coal,* May 1961, p.4.
24. Fabian Group, *Plan for Coal Distribution,* Fabian Society, London, 1956.
25. Ibid., pp.3-4
26. *Coal Quarterly,* Summer 1962.
27. Ministry of Power, *Report of the Court of Inquiry into "Coal Distribution Costs in Great Britain"* (Chairman Robson), Cmnd. 446, HMSO, London, 1958.
28. Select Committee on Nationalised Industries, *Report and Accounts,* HMSO, London, 1958.
29. Ministry of Power, *The Finances of the Coal Industry,* Cmnd. 2805, HMSO London, 1965.

30.   Ibid., p.3.
31.   National Board for Prices and Incomes, *Coal Prices,* Cmnd. 2919, Report No.12, HMSO, London, 1966.
32.   Op. cit.
33.   Op. cit.
34.   *The Financial and Economic Obligations of Nationalised Industries,* Cmnd. 1337, HMSO, London, 1961.
35.   Ibid., p.3.
36.   Ibid., p.3.
37.   Ibid., p.3.
38.   *The Times,* 6 October 1959, p.14.
39.   Op. cit.
40.   Op. cit.
41.   Op. cit.
42.   J.W.House and E.M.Knight, *Pit Closure and the Community,* University of Newcastle, Department of Geography, 1967.
43.   E.M.Knight, *Men Leaving Mining,* University of Newcastle, Department of Geography, 1968.
44.   Department of Employment and Productivity, *Ryhope: A Pit Closes,* HMSO, London, 1970.
45.   Ibid.
46.   *Coal Quarterly,* Winter 1963, p.23.
47.   Department of Employment, *Report of a Court Inquiry into a Dispute*
  B   *Between the National Coal Board and the National Union of Mineworkers* (Chairman Wilberforce), Cmnd. 4903, HMSO, London, 1972.
48.   See Chapter 4.
49.   Quoted by J.Hughes and R.Moore (ed.), *A Special Case?,* Penquin, Harmondsworth, 1972, p.131.
50.   Ibid., p.34.
51.   See, for example, W.Brown, *Piecework Abandoned,* Heinemann, London, 1962.
52.   See M.P.Jackson, *A Critical Analysis of the Ministry of Labour's Method of Classifying the Causes of Stoppages, with Special Reference to Major Stoppages in the Port Transport and Coal Mining Industries Between 1963 and 1966,* University of Hull, unpublished MA thesis, 1971.
53.   National Coal Board, *Annual Report for 1966/7,* NCB, London, 1967, p.37.
54.   See *The Times,* 7 July 1964, p.8.
55.   See Ministry of Labour, *Report of the Committee of Investigation into Differences Between the Yorkshire Winding Enginemen's Association and the Members of the NUM employed by the NCB* (Chairman Wilson), HMSO, London, 1964.
56.   Plan presented to the Minister of Power in November 1959.
57.   *The Times,* 19 November 1959, p.9.

6.   FORGOTTEN POWER

Production and consumption of coal had fallen almost yearly throughout
the 1960s. By the turn of the decade, however, there were signs that the
position might be changing. For the first time for about fifteen years
there was a shortage, not a surplus, of coal in Britain. The position
became quite serious in the winters of 1969-70 and 1970-71. Coal
stocks fell alarmingly; in March 1969 undistributed coal stocks stood
at 25 million tons but by March 1970 they had fallen to just over 14
million tons, and a year later they were at the critical level of just over
6 million tons. Further, there were signs that the cost of coal's main
competitor, oil, was starting to increase. In August 1970 the Trans
Arabian Pipeline, which bypassed the blocked Suez Canal, was broken,
and the Libyan Government took advantage to the increased demand
for Libyan oil [1] to substantially raise its prices. In succeeding months
other oil producers followed Libyan the example. By mid-1971 major
oil users were paying between £6.60 and £9.30 a ton (coal equivalent)
for oil, compared to between £3.30 and £6.00 a ton in June 1970.
The rise in oil prices meant that coal was far more competitive. Much
industrial coal was being sold at below £5.25 a ton (pithead prices) and
Midlands coal averaged around £4.75 a ton in mid-1971.

Reactions to Changing Conditions

The Coal Board reacted by taking immediate measures to slow down
the contraction of the industry. In 1970 total coal production in
Britain at over 142 million tons was only about 8 million tons less than
the previous year, and in 1971 production was increased by 2½ million
tons (see Table 24). The Board also took steps to slow down the rate of
pit closures; in 1970-71, for example, only eight pits were closed and
all of these were closed down as the result of the exhaustion of reserves.
In addition, efforts were made to recruit new miners; at the end of 1970
the Board announced a massive recruiting drive designed to fill another
8,000 jobs.

Early in 1971 these moves were consolidated when the Board
announced its new five-year plan for the industry. NCB Chairman
Robens said that the Board aimed to increase production to 150 million
tons a year, virtually cease pit closures, and increase the size of the
labour force.

We are going for a total output of about 150 million tons a year —
to meet the market we know will be there . . . From now on we are

going to ignore the (1967) White Paper's highly erroneous forecasts . . .
The Board has a statutory responsibility to make the right qualities
of coal available and we intend to do that. It would require a
government directive to change this policy . . . The 'five-year plan'
target is 140 million tons of deep-mined coal a year and about
10 million tons from opencast sites.

It means no more closures, except for the few pits where coal
reserves become exhausted or very severe geological problems
are encountered. Closures for these reasons are likely to be as few
as four to six in 1971 — throughout all the coalfields.

It means job security for the years ahead . . . For the first time
for more than a decade recruitment is running well ahead of
'wastage' — miners retiring or leaving — and manpower is going up.[2]

The government took action in two major areas. First, they took
measures designed to increase the supply of coal. Thus, they permitted
coal imports which had effectively been prohibited throughout the
1960s. In 1971 over 4 million and in 1972 nearly 5 million tons of coal
were imported. They also encouraged the increase in coal production
from open cast mines. In 1971 and 1972 more than 10 million tons
of coal was produced from open cast mines compared to between 6
and 7 million tons throughout the 1960s.

Second, they took measures designed to restrict the consumption
of coal. Thus, in 1971 the Coal Industry Act became law. The Act
contained a wide range of provisions, from those raising the limit of
the Board's accumulated deficit to those dealing with the Board's
subsidiary activities (a controversial plan was outlined whereby the
Secretary of State could order the Board to sell certain of its non-
colliery — and in most cases profitable — activities). In particular,
however, it stopped the practice of subsidies for electricity and gas
undertakings for burning 'extra' coal, which had been introduced after
the recommendations of the 1967 White Paper.[3] In addition, the
government gave permission for the Board's largest consumer, the
electricity industry, to move to a greater dependency on oil. The
Central Electricity Generating Board had for a long while wanted to
move to a greater reliance on oil but had been prevented from doing
so by government policy. This barrier was now lifted. No new coal-
fired generating stations were planned after 1970 and by 1972-3 the
CEGB had cut its coal consumption to little more than 62 million tons,
compared to 70 million tons in 1970, even though in 1972-3 it
generated 13 per cent more electricity than it had three years earlier.
Overall, in fact, home consumption of coal, in response to government
policy and coal shortages, fell quite markedly in the early 1970s, from
154 million tons in 1970, to 138 million tons in 1971, and to less

131

than 121 million tons in 1972.

The miners reacted by pressing for their biggest ever pay increase. They saw the opportunity in the changed market conditions of reversing, or at least halting, the relative decline in miners' earnings that had occurred in the 1960s. At their annual conference in the Isle of Man in July 1970, the NUM approved a resolution calling for wage rises of £5 a week for surfacemen (from £15 to £20), £6 a week for underground workers (from £16 to £22) and between £5 3s. and £2 7s. 6d. for power loading teams (from £24 17s. − £27 12s. 6d. to £30). The Coal Board made an offer which was worth slightly less than half of the miners' claim. They proposed wage rises ranging from £2 10s. for surface and underground workers to £1 17s. 6d. for power-loading teams.

The NUM rejected the Board's offer and determined to ballot their membership on a national strike. Under the union's rules an official national strike could only be called if such a move was approved in a ballot of the membership by at least two-thirds of those voting. In the event the ballot, held in early October, produced a majority in favour of strike action (55½ per cent) but not the two-thirds majority required by the rules.

Following the ballot negotiations were reopened between the NCB and the NUM. The union had not obtained a sufficient majority to enable it to call a national strike, yet the ballot had shown that there was considerable dissatisfaction with the Board's offer, dissatisfaction which was later to manifest itself in unofficial strike action. The NCB, therefore, agreed to raise the original offer. Pay rises of £3 a week for surface and underground workers and £2 7s. 6d. for power-loading teams were suggested. It was estimated that the new offer would cost about £39 million. This offer was put to the membership through another national ballot. This time over 65 per cent of those who voted declared for acceptance (only Scotland, South Wales and Kent voted against it).

Industrial Dispute of 1971-2

The 1970 dispute, however, proved to be no more than a curtain-raiser for the widespread industrial action that occurred the following year. The scene for this much more serious dispute was set by the 1971 annual conference of the miners' union held in July at Aberdeen. At that conference two crucial decisions were taken. First, the conference decided to reduce the majority necessary in a ballot before the union could call a national strike. Two resolutions were put forward, one which would have meant that only a simple majority of those voting would have been necessary, the other which fixed the necessary majority at 55 per cent. In the event both resolutions were carried but

132

only the second one obtained the two-thirds majority necessary to change the union's constitution. Second, the conference approved a resolution calling for a major pay claim, backed up if necessary by industrial action. The resolution read as follows:

> This conference of the National Union of Mineworkers calls upon the National Executive Committee to submit a claim for a substantial increase in wages of all its members and to seek to establish a minimum wage on the surface of £26 a week, a minimum wage of £28 per week underground and a minimum NPLA rate of £35 per week. Further it should renegotiate all differentials in order to establish a realistic measure of job values, and seek a simultaneous date for the operation of wage settlements for all sections coming under the Coal Mining Industry Wages and Conditions. In the event of unsatisfactory response from the National Coal Board the National Executive Committee to consult the membership with regard to various forms of industrial action.[4]

The Coal Board's response was to offer to increase miners' wages by rather less than they had done a year earlier. They offered wage increases ranging from £1.80 to £1.70 a week. This response was undoubtedly influenced (although the precise nature of the influence was a matter of dispute) by government economic policy. One of the central aims of government economic policy was to reduce the level of inflation; and they believed that in order to do so they had to reduce the level of wage settlements. As a result the government suggested a norm for wage settlements of about 7-8 per cent. The NUM rejected the offer and embarked on a programme of prolonged industrial action which included an overtime ban and a national strike.

In one sense the miners' reaction might have been expected from a review of events over the previous few years. There had been signs of growing restlessness of rank and file members between 1969 and 1971. In 1969 and 1970, for example, major unofficial strikes occurred, which led to the loss of about one million working days each, and in 1971, although the NUM did not authorise official action, there was a rash of unofficial strikes. The reasons for this restlessness are many but include at least three factors. First, the miners had witnessed a massive contraction of their industry over the previous decade; coal production, employment and the number of pits in operation had all fallen dramatically. Second, the sacrifices of the miners over the previous decade had not led to any improvement in the position of those who were left in the industry. In many other industries contraction in size has been accepted by the workers because they have been told that those who remained would be better paid and have secure jobs. In the coal-mining industry quite the opposite seemed to have occurred. Those who remained seemed far worse off, certainly in terms of earnings.

The decision to push their claim to industrial action was also undoubtedly influenced by a third factor; the view that the industry might, even if only temporarily, be enjoying an improvement in its market position. Few miners believed that the improvement was destined to last and they remained pessimistic about the long-term future, but most felt that they should take advantage of it while they could. Certainly the NUM leaders used the improvement in the industry's market position as a major argument in favour of their wage claim. Thus General Secretary Daly said,

> Coal production is to be expanded from 140 to 150 million tons a year, but it will not be possible to attract and keep the young and skilled manpower needed to achieve this unless the Board can pay the kind of rates we are demanding.[5]

The overtime ban began on 1 November 1971. Its initial impact was a matter of dispute. On the one hand, the Coal Board's Deputy Chairman, Sheppard, claimed that the ban had been relatively ineffective.

> When the NUM began their action at the beginning of November, I was warned that production would be . . . reduced . . . What has happened?
>
> 1. NCB stocks have increased.
> 2. Stocks held by coal users are still at record level.
> 3. The pits have been saved the heavy vosts that would otherwise have been involved in vast stocking.
> 4. Hundreds of thousands of pounds have been 'saved' which would otherwise have been paid for normal overtime.[6]

On the other hand, Gormley, NUM President, claimed that the NCB was deliberately underplaying the effects of the ban.

> I am more than satisfied with the tremendous response to our appeal for an overtime ban. I realise that there have been one or two local difficulties, but in general the ban has been operated successfully successfully. And it is, in my opinion, having a greater effect than the Board estimated or are willing to disclose.[7]

Whatever the true position it soon became clear that the union was determined to increase the scale of the conflict if a better offer was not forthcoming. At the end of November it conducted a ballot of the membership on the issue of a national strike in support of the wage claim. In the ballot the union obtained the backing of a majority of its members; 59 per cent voted in favour of strike action. Despite the relatively narrow margin of support the NUM executive decided unanimously to call a national strike the following month (union rules required one month's notice of an official strike).

For a while, though, it seemed as if the strike might be averted. Two factors supported this view. First, there was a fairly widespread belief that the miners (and their unions) did not want a strike and could not really afford one. Thus,

> The miners' executive voted unanimously on Thursday to call a strike from January 9th, but took the precaution of announcing at the same time that it was reopening talks with the National Coal Board. The miners did not look in a particularly strong position. No worker wants to hear strike talk so close to Christmas, and there is plenty of coal in stock to ride out a lengthy strike after it. There is at least ten weeks supply at full winter rates of consumption already at the power stations, which is enough to see them through to the end of March.[8]

This was allied to the feeling that a national coal miners' strike would inevitably lead to a further contraction in the size of the industry. Thus, *The Economist* argued:

> It is two years since a pit was closed for any other reason than that it was worked out. The high output, high profit coal-fields of South Yorkshire and Staffordshire should have a future for as far ahead as anyone can forecast. But this leaves out in the cold all those pits in Scotland, Wales and such English fringe fields as Durham . . . the Coal Board knows that a strike will now only bring forward a new programme of closures, making a considerable number of young, fit men redundant. The next pit shutdown plan cannot include the re-employment opportunities which the Coal Board was able to offer in the past by transferring miners to areas where labour was short. The miners caught in areas of already high unemployment have little but the scrapheap to face.[9]

Second, there were the moves, referred to in one of the above quotations, to reopen talks between the NCB and the NUM. On 21 December the NCB presented a revised pay offer to the union. The package offered pay increases of £2 a week on the minimum surface rate (bringing it up to £20), an extra £1.90 for all other grades, an extra week's holiday (from 1 November 1972) and talks on a bonus scheme linked to productivity, which was to be worked out in time for the next wages settlement. The NCB, however, emphasised that this new offer was as 'far as they could possibly go'; they were unable to go beyond the bounds of government pay policy and treat the miners as a 'special case'. As a result, when the union executive rejected the offer, the previous optimism about the strike being averted began to disappear.

The strike began on 9 January, but for a while (for the greater part of January), in many ways like the overtime ban that preceded it, caused little concern. The concern that was shown centred on the 'social problems' that the strike caused rather than any threat the strike might pose to the economy of the nation'. Concern was expressed, for example, about the closure of schools (schools in the North and West were particularly badly affected) and the effect the strike might have on the elderly. On 23 January the organisation 'Age Concern' began a national campaign to help old people affected by the lack of coal and heating.

The miners' union seemed to be facing problems on two fronts. First, there was evidence that some rank and file members were unhappy about the way it was conducting the dispute; many felt that the union was not pursuing the dispute vigorously enough. Thus, for example, many miners ignored the NUM's directive that members hould continue to provide safety cover at the pits; about half of the pits were without safety cover when the strike started. Second, the union did not receive the degree of support they had hoped for from the rest of the trade union movement. On 10 January, for example, the TUC Finance and General Purpose Committee turned down an NUM request for a 'round table' conference of unions to generate support for the strike. The TUC, instead, restricted itself to urging members not to cross picket lines. The miners were clearly shaken by the TUC's response. NUM President Joe Gormley, said after the meeting:

> I am extremely disappointed to know that they (the TUC) did not consider this (the miners' strike) serious enough to get together to form some concerted ideas. I would have thought this was one time when the TUC could have shown itself to be united.
> I would not say we are on our own yet. I know some of the individual unions will react favourably, but it would have been better if action had been centralised.[10]

The government for their part seemed to be content to let the strike take its course, at least for a while. Despite the miners' overtime ban coal stocks were high; it was estimated at the beginning of the strike that coal stocks at power stations were sufficient to last for 8·8 weeks of winter use. The government's main efforts seemed to be directed towards persuading the public that their stand was just and persuading the miners that they could not win. Central to their strategy on this latter point was the emphasis placed on the by then well-publicised argument that prolonged strike action by the miners could only lead to a more rapid contraction of the mining industry. Thus, Davies, Secretary of State for Trade and Industry, warned the miners in a speech on the second day of the strike, that 'continued strike action would do irreparable damage from which the industry might never recover.[11] He said that, 'there were other fuels waiting in the wings to take over from coal the nation's power requirements.'[12]

The position changed, however, towards the end of January and the strike began to be taken much more seriously. No longer was the strike merely causing concern because of its 'social consequences'; it was now seen as potentially damaging to the whole of British industry. Two factors seem to have been mainly responsible for this change. First, there was danger for a while that the miners might be joined on strike by the power workers. They threatened to take such action in pursuit of their own wage claim, and although on 7 February they accepted a revised offer (giving them wage increases averaging a little over 7 per cent) their attitude clearly awakened many people to the potential dangers of the situation. Second, and in the event more crucially, it became clear that the miners' picketing was having a considerable effect. The government, in their assessment of the likely impact of the strike, had merely taken the amount of coal lost through the cessation of mining into account; they had not allowed for the fact that it might prove impossible to make coal stocks available to the power stations because of the miners' picketing. The result was that the power stations began to feel the effects of the coal shortage much sooner than had been expected. By early February the CEGB was forced to make widespread voltage reductions and they announced that power cuts had only been avoided because of the 'mild weather'.

The worsening in the supply position was undoubtedly one of the factors that persuaded the NCB (presumably with government support) to reopen negotiations with the NUM. An improved offer was eventually made which would have given miners rises of between £2.75 and £3.50 a week (compared to the £1.90 to £2 of the previous offer). The NUM executive, however, decided to reject the new offer and continue with the strike.

The NUM's rejection of the revised offer was the signal for yet a further escalation of the dispute. The optimism that the nation could 'ride out' the effects of a miners' strike was by now completely dispelled; the power situation was becoming desperate. Forty-four power stations were working well below capacity and the miners' picketing was preventing not only coal but also essential raw materials, such as oil and hydrogen, getting through to the power stations. The government had already proclaimed a State of Emergency which had led to rota power cuts of up to four hours a day throughout the country. On 11 February the government announced even more stringent measures to restrict the consumption of electricity. Industrial consumers with an estimated demand of 100 kilowatts or more were required to restrict their use of electricity to three days a week, excepting Sunday; larger industrial consumers and continuous process industries were required to cut consumption by 50 per cent; the use of electricity for heating offices, shops, public halls, catering establishments and premises used for recreation, entertainment and sport was banned. Prime Minister Ted Heath spelled out the likely impact of the restrictions:

> From next Monday (14 February) a great part of British industry will be on half time. Millions of men and women will be put out of work as their factories close down. Some jobs will be permanently at risk.[13]

This prophecy was borne out by events. The *Guardian* reviewed the effects of the first day of restrictions in the following way:

> Thousands of workers in car, engineering and textile firms, mostly in the North and Midlands, were laid off yesterday as British industry felt the full impact of the power crisis . . . Nine manufacturing plants at the Lucas components group were idle in Birmingham, with 13,000 laid off. Lay-offs are to be spread among the rest of the firm's 21,000 labour force as the crisis continues. Other lay-offs among car manufacturers were at Jaguar, Coventry (3,500), Triumph, Coventry (2,500) and Rover (2,000) . . . General engineering suffered similarly. Half of the 4,000 hourly-paid workers at the big GEC-AEI factory at Trafford Park, Manchester, were laid off for the day; the other half will stay at home today as the company operates a day-on day-off rota . . . The cotton and man-made fibres section of the industry . . . laid off thousands of workers . . . Many firms have serious doubts about their ability to work economically under the government's time-table, and some have already laid off their employees for the duration of the crisis . . . Among other firms which announced lay-offs yesterday were Plessey, Merseyside: 5,000 today; Rubery Owen, Staffordshire: 4,000 yesterday; Rowntree-Mackintosh, York: 3,000 idle tomorrow and Thursday; Montague Burton, Leeds: 8,000 until tomorrow.[14]

The *Sunday Times* forecast even worse to come:

> It is now only a matter of a few weeks — perhaps three or four —
> before the present phase of power cuts is followed by total
> disconnections of electricity supply from a large number of Britain's
> industrial and domestic consumers.[15]

On 16 February the Department of Employment announced that as a
result of the restrictions a total of 1·2 million workers had been 'laid-off'.
There were even fears that the restrictions would lead to food shortages.

> Milk, eggs, and bread may become particularly vulnerable as the
> power cuts take their toll. In a number of cases milk rationing has
> begun.[16]

It was against such a background that the Wilberforce Inquiry had to
conduct its investigations. The government had set up the Court of
Inquiry to examine the causes and circumstances of the dispute when it
had announced its emergency restrictions. The Court of Inquiry was to
prove to be of crucial importance. It provided the miners with an
important public platform on which to put their case and in so doing
enabled them to consolidate the public sympathy they had attracted
during the strike. The inquiry was also, through its report, to provide a
basis for the eventual settlement of the dispute.

Evidence was given to the inquiry by a range of individuals and
institutions. As might be expected they expressed a variety of opinions
and proposed many different courses of action. Probably the most
interesting and important aspect of the evidence, however, was the
support given to the miners' case by a number of 'outside witnesses'.
These 'outside witnesses' were able to present a considerable body of
detailed evidence to support the miners' case, and the Court was clearly
impressed. Central among this group was Hughes, Director of the Trade
Union Research Unit and Vice-Principal of Ruskin College, Oxford.
He produced evidence to show that miners' earnings had fallen well
behind those of other workers. Thus, Hughes argued that whereas in
April 1967 earnings in the coal-mining industry had been 7 per cent
greater than those in manufacturing industry, in April 1971 they were
7 per cent less (and, as Hughes argued, the latter date was a particularly
bad one for the manufacturing industries and therefore the figures
probably did not show the full extent of the decline in miners' earnings).
One of the main reasons for the decline in miners' earnings, particularly
the earnings of faceworkers, he argued, had been the reduction in the
extent of the use of piece-work by the industry. In 1968 (September)
piecework had accounted for 11·5 of all miners' earnings whereas in

139

1970 (April) it had accounted for only 2·3 per cent. The miners had, of course, supported the abolition of piecework through the National Power Loading Agreement. Their complaint was not that piecework had been abolished but that basic wage rates had not been increased sufficiently to compensate.

Other 'outside witnesses' included Meacher, a Labour Member of Parliament and ex-university lecturer, and Clegg, Professor of Industrial Relations at the University of Warwick. Meacher presented evidence on what has been called the 'poverty-trap'. Briefly he argued that when income tax and means-tested benefits were taken into consideration miners would not benefit from the NCB's offer as much as might at first sight appear. Many miners, he argued, would gain little in real terms from the NCB's offer and some might actually be worse off. His conclusion was that in order to avoid the 'poverty trap' lower-paid workers needed to be offered very substantial increases in wages, certainly increases far more substantial than those being offered by the NCB.

Clegg, in his evidence, discussed the operation of pay and prices policies. He argued that the need to control inflation should not preclude certain groups of workers, in special circumstances, being given exceptional wage rises. He said:

> our industrial relations system is in need of some permanent provision whereby consideration can be given to the case for exceptional pay increases in industries and undertakings in a situation such as that of the coal-mining industry over the last year or so.[17]

The inquiry's report[18] broadly supported the union's case. Specifically it recognised that coal-miners' earnings had fallen behind those of other workers. It showed (see Table 25) that whereas coal-miners had been third in the earnings league table in 1960 and 1965, by 1970 they had fallen to twelfth and by 1971 they were still only ninth. (The improvement in the 1971 position was partly the result of the fact that earnings in manufacturing in October 1971 were depressed below their normal level because of a temporary recession in business activity.) The Court did not argue against any changes to the relative earnings of different occupations but did argue that the extent of the changes that had taken place *vis-a-vis* the coal-miner could not be justified.

> Some changes (in the relative wage positions of different industries) are necessary from time to time as economic circumstances, technology and other factors alter. In our view,

however, the fall in the ranking of coal-mining pay has been quite unwarranted by any such changes.[19]

The report recognised that the coal-miners' claim could not be treated in isolation. Problems of inflation were kept uppermost in the Court's eyes:

we cannot shut our eyes to the continuing danger of inflation, which presents the most serious threat to the standards of living of everyone, including the miners themselves.[20]

The Court felt, nevertheless, that the miners were a 'special case' and deserved to be given special treatment. They put forward five main reasons to support this argument. First, surface workers who were on the minimum rate of £18 a week were among the 'lower-paid'. They lived in communities where there was little opportunity for alternative work and many of them had been forced to accept work on the surface because of sickness and injury suffered in the service of the mines. Second, the work of the men who laboured underground was particularly arduous and dangerous.

Other occupations have their dangers and inconveniences, but we know of none in which there is such a combination of danger, health hazard, discomfort in working conditions, social inconvenience and community isolation.[21]

Third, there were the special problems of inconvenience and danger suffered by the miners who worked at the coal face. They had to endure all of the difficulties associated with general underground work, and more. Fourth, the Court cited the 'exceptional' co-operation shown by the miners in the past to reform the industry's wage structure: a reform which had been essential to enable the industry to carry through its plans for modernisation. Fifth, the court noted that shift payments were minimal or non-existent in the industry. The Court concluded that

for these reasons which are exceptional and do not apply in industry generally . . . the mine-workers at this particular time have a just case for special treatment.[22]

The specific recommendations made by the Court were for wage rises of £5 a week for surface workers (making a minimum rate of £23) of £6 a week for underground workers (making a minimum rate of £25) and of £4.50 a week for power-loading teams (making a minimum rate of £34.50). The Court recommneded that the increases should be back-dated to 1 November 1971 and that the new rates should continue in force for sixteen months, until February 28, 1973. The Court did not make any firm recommendations on two other aspects of the NUM's claim, those concerning the age at which the adult rate should be paid and holidays, but it did make recommendations on one further matter, that of productivity. The Court supported the moves that had been made to reduce the use of piecework in the industry. It argued, however, that this should not be taken to imply that they were opposed to all productivity incentives and it suggested that the NUM and the NCB should enter into further discussions (to be completed by the end of March 1972) in an attempt to devise a suitable productivity bonus scheme.

The initial reaction of the NUM was to reject the Wilberforce recommendations. Thus, the national executive committee decided by 13 votes to 12, on 18 February, that it could not support a settlement based on these recommendations. Later the same evening, however, they responded to an invitation to meet the Prime Minister to discuss the matter. At 1.00 a.m. on 19 February the NUM executive agreed to recommend that their members return to work. They had not obtained the £1 a week increase on the Wilberforce terms for surface workers that they had wanted but they had succeeded in gaining a number of concessions[23] not included in the Wilberforce Report.

The NUM executive, of course, could only make a recommendation that the miners return to work. The actual decision to call off the strike was in the hands of the rank and file members. They voted on the issue on Wednesday 23 February and the result of the ballot (supervised by the Electoral Reform Society) was announced on the following Friday: 217,000 miners had voted in the ballot and over 96 per cent of these had voted in favour of a resumption of work. There was a full resumption of work on Monday 28 February.

It is, of course, difficult, if not impossible, to explain fully why events took the course they did. It might be worth-while, however, pointing to three factors which seem to have had a particularly important impact.

The first concerns the question of public support. Public opinion was no doubt of considerable importance, if only in a negative fashion. As Richardson and Jacobs pointed out, 'only when public opinion had swung absolutely against the miners would the government feel free to bring the troops in'[24] And with the miners so determined in their stand, bringing the troops in to move stocks of coal, and possibly to help import coal, seemed to be one of the few options open to the government. Yet, it is clear that public opinion did not swing decisively against the miners. Certainly there were attacks on the miners from

142

many quarters. An editorial in the *Sunday Times* (a paper whose support the miners might have had some hopes of gaining) on 13 February said, for example:

> The miners are united and determined, with the rare seriousness of people who have not struck since 1926 and are supported by a majority ballot. But they are also now unreasonable.[25]

Feature writers in the same paper went further than this. R. Butt, for example, in an article in the same edition suggested that the government might be justified in calling a general election on the issue in order to demonstrate that they had public support and E. Jacobs argued that the strike was not really about the miners' 'so-called' grievances but had been caused by the poor quality of leadership in the miners' union.

Yet, despite such comments, the miners did retain a good deal of public support; and support came from some of the most unexpected sources. For example, in an editorial on 19 February (the day after the Wilberforce Inquiry reported) the *Daily Express* commented on the justice of the miners' case:

> Let there be no mistake. The nation wanted the miners to get a fair increase. They have been treated as a 'special case'. And rightly so. The terms of the settlement were virtually those suggested by this newspaper.[26]

One of the reasons why the miners received so much support was no doubt the strength of their case. Yet there is also no doubt that they maintained a public relations campaign of greater sophistication than that attempted by most other unions in previous disputes. The campaign which included a series of half-page advertisements in national newspapers at a cost of £45,000 was certainly more professional than one had come to expect from many trade unions.

Related to the question of public opinion is the second factor that might be mentioned, that of the solidarity of the miners and the nature of the mining community. The miners were not subjected, as were the 'power workers' in 1970, for example, to the full force of what adverse public opinion there was, simply because by and large they live in separate communities. This point was specifically referred to in an article in the *Guardian* on 15 February.

> Picketing miners who are slowly starving the power stations into closure are not likely to be deterred by any charges of 'holding the country to ransom' or warnings about backlash from an enraged public . . . Because most miners live in pit villages or on Coal Board

estates they do not feel as individually vulnerable to public outrage as did, say, the power workers. Two years ago the power workers were almost stampeded back to work because they were widely scattered among a public intent on lashing back with violence, insults or contempt. Many people have railed against the social undesirability of the 'industrial ghettos' in which the miners live, but for once the 'ghetto' is serving a useful purpose as an industrial refuge.[27]

Another result of the nature of the mining community is the feeling of 'solidarity' this can engender. Miners throughout the world have for a long while been seen as a particularly strike-prone group of workers. When Kerr and Siegel attempted an international comparison of the inter-industry prospensity to strike[28] they found that miners and dock workers (marine and longshoremen to use their terminology) topped their list. One of the main reasons suggested by Kerr and Siegel why both of these groups of workers were involved in a comparatively large number of strikes was the fact that they tended to live in single industry communities and that this, in turn, led to a greater feeling of solidarity and resolve than amongst other workers.

Such reasoning would certainly seem to be applicable to this strike. It was reported, for example, that in the East Kent mining village of Betleshanger the strike brought 'the village closer together than it has ever been since the war'.[29] Similarly, it was reported that in many mining towns the strike was being partly financed by local tradesmen who either reduced their prices for the duration of the strike (this happened, for example, at Mansfield, when the price of a loaf of bread was reduced by the local baker from 10p to 6p — nearby at Allerton three out of every four men work in the mines) or allowed miners credit.

The two factors mentioned so far help to explain why the government felt unable to take decisive action to deal with the strike and why the miners were able to endure for such a long while, but they do not really explain why the government felt unable to prevail upon the Coal Board to resist the strike for a longer period of time. To gain an understanding of this factor we really have to return to a consideration of government economic policy.

The aim of the government was essentially to reduce the level of wage settlements. They wanted to do so, explicitly at least, in order to control inflation. It was, therefore, seen to be essential to oppose the miners' strike. Yet the paradox as far as the government was concerned was that the longer the strike lasted and the longer they resisted the miners' demands, then the greater the possibility of inflation resulting merely from the strike itself. The reduced working week, for example, was contributing to inflation; it meant idle plant and increasing costs. Commenting during the strike Harris said:

> They (the Government) must catch up with their thinking now, because it is becoming clear that if the strike goes on much longer there will be some real disruption. If we get disruption it could be inflationary. That is the real trouble.[30]

Added to this there was the problem of 'confidence'. One of the other main aims of the government's economic strategy was to revive confidence in the economy so as to bring about an upturn in industrial investment. The miners' strike, particularly as its effects became more serious, did nothing to revive confidence. Thus it was said:

> A Chancellor whose economic strategy is built on the need to build up confidence – the confidence that will get manufacturers investing and consumers spending – needs the power cuts as much as Harold Wilson needed the seamen's strike.[31]

Further, it should be remembered that the cost of the miners' pay award, as agreed, was estimated at about £90 million. This was almost precisely the figure that the NCB estimated they had lost as a direct result of the miners' strike.

1972-3 Wages Dispute

Within a matter of months of the settlement of the 1971-2 wages dispute the prospect of a further confrontation over wages loomed large on the horizon. At their annual conference in July 1972 the NUM passed a resolution demanding further wage increases. The resolution, put forward by Derbyshire and Kent areas, argued that modifications needed to be made to the Wilberforce agreement to take account of the rise in the cost of living. Minimum rates of £30 a week for surfacemen, £32 a week for underground workers and £40 a week for power-loading teams were demanded. The resolution concluded,

> In the event of an unsatisfactory offer from the NCB, the NEC should consult the members with the view to taking various forms of industrial action.[32]

In the event the fear of further confrontation receded for the time being. The Coal Board initially made an offer based on the government norm of £1 plus 4 per cent, although after the NUM rejected this offer, certain improvements were made, particularly in the area of fringe benefits. The final package gave all miners wage increases of £2.29 a

145

week; doubled the miners' pension, from £1.50 to £3 a week (at no extra cost to the miner); restored the value of allowances for miners living in smokeless areas to the 1961 level; and consolidated the holiday provisions, so that in future a miner would receive a total of three weeks annual holiday, plus seven rest days and six statutory holidays. The NCB's offer was submitted to the miners and was accepted in a ballot by 143,006 votes to 82,631.

Changing Fortunes

Many feared and warned that the disputes of the early 1970s would cause lasting damage to the industry. Apart from the physical damage to the mines and mine equipment it was believed that the disputes might result in the industry losing, possibly permanently, valuable markets and in further massive financial losses for the Board. Initially some of these fears seemed to have been well-founded. In 1972, for example, domestic coal consumption, at about 121 million tons, was 18 million tons less than it had been the previous year, and the industry's main customer, the CEGB, had started their switch away from coal towards oil. In 1972 oil provided over 48 per cent of the fuel burnt at power stations in Britain. Further, in 1971-2, despite a government grant of £100 million, the industry made a loss of £91 million (compared to a surplus of £½ million the previous year).

In 1972 the government introduced a Bill into Parliament to give aid to the once again ailing coal industry. The Bill which became law in March 1973[33] reversed some of the provisions of the 1971 Coal Industry Act. In particular, encouragement was again to be given to the electricity industry to burn more coal. In April 1973 the CEGB and the Coal Board concluded an agreement for the electricity industry to increase its coal consumption to about 65 million tons a year. Financial assistance was provided for the CEGB, under the terms of the 1973 Act, for all coal burnt by power stations in excess of 58 million tons. The 1973 Act also provided help for the coal-mining industry in a number of other ways. First, under the Act, the government agreed to give direct support to the Board to offset the costs of carrying coal and coke stocks and providing higher pensions. The support was limited to a maximum of £120 million a year. Second, the Act provided for an increase and extension of the redundancy provisions of earlier legislation. Third, financial liabilities of the Board, totalling £475 million, were 'written off'.

In mid-1973, however, the coal-mining industry's fortunes once again took a dramatic turn. This change in the industry's fortunes occurred largely because of the effects of decisions taken by oil producing countries to increase the price of oil and to limit its supply. At a meeting of the Organisation of Arab Petroleum Exporting

Countries held in Kuwait in October 1973, a decision was taken to reduce the level of oil production by 5 per cent a month from the September 1973 level,

> until Israeli withdrawal is completed from the whole Arab territories occupied in June 1967 and the legal rights of the Palestinian people are recognised.[34]

Further, exports to certain countries, notably Holland and the USA, were cut altogether. The limitation of supplies was only part of the 'oil weapon' the Arab countries determined to use to put pressure on nations 'friendly' to Israel to persuade her to withdraw from occupied territories. The other part of the 'oil weapon' concerned price. Exporting countries had raised the price of oil substantially in 1970, 1971 and 1972; in 1973 the price was raised once again, and by a far greater amount. Most prices were doubled in October 1973 and then doubled again from 1 January 1974 (see Table 26). Thus, the price of Arabian light oil was increased from 1.479/3.370 dollars a barrel in January 1972 to 11.651/13.647 dollars a barrel in January 1974. It should also be noted that the prices quoted in Table 26 refer to the 'posted' prices of oil. Often the market price is different from the posted price and frequently in 1973 and early 1974 consumers were paying well above the posted prices for oil.

Despite the developments in nuclear power and natural gas, oil was still coal's main competitor. The decision to limit the supply and substantially raise the price of oil had a tremendous effect on the coal-mining industry; overnight almost attitudes towards the industry changed. A restriction on the supply of Arabian oil affected Britain markedly. Although Britain did not obtain all of her oil from Arabian countries she certainly obtained the vast bulk of it from them, and the cutbacks in Arab exports had a serious effect. Possibly even more important than the physical effect of the cutbacks, however, was the effect that it had on attitudes towards different sources of energy. Far greater importance was in future to be attached to 'secure' energy supplies. The Arab oil restrictions had given a great boost to the arguments that Britain should not allow her coal industry to be 'run down' because too great a dependence on oil would make her economically vulnerable to action by other countries.

The raising of oil prices also meant that coal became far more competitive. The way in which this happened can probably best be illustrated by reference to the coal-mining industry's major consumer, the CEGB. In 1974 the CEGB was paying about 6p or 7p a therm for oil, whereas the average cost of coal was only about 3p a therm. Even when account is taken of the government subsidy to the coal industry and the excise duty on imported oil, coal in early 1974 was still substantially cheaper than oil.

1973-4 Industrial Dispute

In July 1973 the NUM at their annual conference, held in Inverness, adopted a motion proposed by M.McGahey, on behalf of the Scottish area, calling for substantial wage increases; nearly £10 a week for surface workers (minimum rate to be increased from £25·29 to £35) about £13 for underground workers (minimum rate to be increased from £27·29 to £40) and about £8·50 for faceworkers (minimum rate to be increased from £36·79 to £45). It was estimated that wage rises of this order would cost the Coal Board £138 million.

At that time the government was still in the process of deciding its strategy for 'Phase Three' of its 'Counter Inflation Programme'[35] and there were reports that the government was keen to ensure that its new policy would provide sufficient scope for the Coal Board and the miners to settle the current pay claim without resort to industrial action. When the Consultative Document on Phase Three was published in October 1973 it contained a provision which was to enable special payments for what became known as 'anti-social hours'. The government believed that such a provision would be particularly advantageous to the coal-mining industry. Government figures showed that coal-mining was at that time the only major industry where shifts were worked but no substantial payment was made for such hours of work. Thus, the Department of Employment earnings survey[36] provided data to show that in April 1973 only 3·2 per cent of miners' earnings were accounted for by shift-work payments (a comparison might be made with the percentage for other industrial groups, for example, printing and publishing, where the percentage was 17·8 and metal manufacture where the percentage was 17·3).

When the Phase Three consultative document was published in October 1973 the Coal Board immediately made an offer to the NUM. The package came within the Phase Three guidelines but the Board felt nevertheless that it was likely to prove to be acceptable. The package contained a variety of elements; a basic addition to wage rates of 7 per cent (£2·25 a week for surface workers and £2·50 a week for underground workers); an increase in the night shift premium from 2½p to 17p an hour (effectively an extra £6·80 a week when men worked the night shift); improvements in holiday pay and differentials made with the help of a 1 per cent flexibility margin permitted under Phase Three; and a threshold agreement which would give miners another 40p a week if the retail price index rose by more than seven points (for each subsequent one point rise they would get an extra 40p a week). In addition, the Board proposed to start talks with the union on the

148

introduction of a 'productivity deal' which would add another 3½ per cent to the package. It was estimated that excluding the productivity deal the package would cost £42 million a year or an extra 13 per cent of the current wages bill.

Initially, many commentators, like the Board, felt that the offer was likely to prove acceptable to the NUM. Thus a report in *The Times* said:

> The threat of a miners' strike this winter receded yesterday after a £42 million Phase Three offer, which would mean wage rises of up to £9 a week for half the industry's labour force. Leaders of the National Union of Mineworkers went through a ritual rejection offer, and called a delegate conference to decide whether the miners should be balloted on strike action; but the skilfully-designed package is expected to prove acceptable in the coal-fields.[37]

One of the reasons why many people expected the offer to be accepted was not merely its size, but also its detailed composition. Under the terms of the offer a high proportion of the industry's face-workers would have received wage rises well above the average; some would have approached the £9 a week suggested in the above quotation.

In fact, such optimism proved to be ill-founded. Not only did the NUM executive reject the offer but on 8 November it unanimously decided to call a complete overtime ban. There are a number of reasons why the executive defied earlier predictions and took such action. One reason was that the offer, in some ways, was just too good as an opening gambit; it left no room for improvement and thus no room for NUM negotiators to 'force' further concessions out of the Board. The Cumberland miners' leader, Rowe, said of the offer,

> It was an offence to everybody's negotiating instincts. There was scarcely any room for manoeuvre.[38]

A second reason was that the offer was divisive. Although it offered a number of miners wage rises of up to £9 a week, it only offered others £2·30 a week. The union had for some time argued strongly against any moves to significantly widen differentials in the industry and was particularly concerned to protect and improve the position of the industry's lower-paid workers. The package specifically went against this aim.

A third, and probably more important reason, was the 'oil crisis'. Between the framing of the union's original demand in July and the rejection of the Board's offer in November a crucial change had taken place in the position of the coal-mining industry. The industry, because

149

of the cuts in oil deliveries and the increase in oil prices, had been recognised as essential to the economic well-being of the nation. Even the 'moderate' members of the executive who earlier might have voted to accept the Board's offer were now no longer willing or indeed able to do so. The ordinary miner saw that the union had an opportunity to press home its advantage and gain significant benefits for its members and he was unlikely to look kindly on any member of the executive who did not go along with such a move.

Announcing the decision to call an overtime ban Gormley, NUM President, said that he felt the action would begin to bite fairly quickly.

> The decision was taken in the knowledge that this will cause a lot of disruption. It is my estimate that if the overtime ban is properly applied, there will be very few pits working after the first week. The crunch will come next weekend.[39]

Despite the fact that coal stocks on 12 November (the start of the overtime ban) stood at 35 million tons, 7 million tons more than they had been at the start of the 1972 strike, and the power stations themselves held 17 million tons of coal stocks (enough to keep power stations supplied for about ten weeks) the commencement of the overtime ban immediately brought an 'emergency' reaction from the Electricity Board and the Government. On 12 November the government declared a State of Emergency. A whole range of regulations were introduced designed to save power; display advertising and floodlighting were banned, government buildings were ordered to cut fuel consumption by a tenth and the general public were urged to make voluntary reductions in their use of power.

There was a great deal of public discussion at the time about the extent to which such moves were really necessary and the miners' leaders openly claimed that the government had over-reacted. On the other hand the government and the CEGB argued that the position was more serious than many people realised; fuel supplies were not only threatened as a result of the shortage of coal and oil but they were also threatened because the power engineers were enforcing an overtime ban in pursuit of their own wage claim. The co-operation of the power engineers was essential if the CEGB was to be able to make 'selective' electricity cuts to save power, as it had during the last coal dispute, but the power engineers' industrial action meant that such co-operation would not be forthcoming on this occasion.

In the meantime discussions had been reopened between the NUM and the NCB on ways of improving the package offer. At a meeting on 21 November the NCB suggested that they might be able to go some way towards meeting the union's argument that the offer did not give sufficient to the lower-paid workers. The idea was to deduct £1.10 from the pay of night workers and put the money saved into a fund which could be used to raise the wages of lower-paid men by up to £3 a week. Such a manoeuvre had been accepted by the Pay Board as within the Phase Three provisions. The NCB's suggestion was unanimously rejected by the NUM national executive.

Later in the month another idea was floated for improving the Coal Board's package offer. This idea involved the payment of miners for time spent at the pit before and after work. If this could be paid for at overtime rates then the size of the Coal Board's offer could be substantially increased. The idea is reputed to have originated from the NUM executive but it was first publicly mooted by Harold Wilson, the Leader of the Opposition. The idea was almost immediately rejected by the government.

Early in December the power position worsened considerably. Two factors had caused the position to deteriorate. First, the miners. overtime ban had really started to 'bite' and stocks of coal at power stations had declined alarmingly. The CEGB was reported to have

> warned the government that coal stocks at power stations are now dwindling so rapidly that further drastic cuts must be imposed within a matter of days.
>
> Unless this is done electricity officials fear that they will not be in a position to maintain the 25 per cent of output which must be guaranteed to keep basic services going even in 'siege economy' conditions.[40]

Second, the situation was complicated by the decision of the train drivers' union, ASLEF, to take industrial action in support of their own pay claim. The bulk of coal was still transported by rail. It was feared, therefore, that even the coal which was mined might not be able to be transported from the pithead to the power stations.

As a result, on 5 December the government introduced a range of new restrictions. For example, street lighting was severely curbed and the temperature in commercial premises and offices was restricted to 63 °F. More crucially, a week later, the government announced that it had decided, in order to save power, that the bulk of industry and commerce would be restricted to a three-day working week. The three-day week was to commence on 30 December but in the meantime industry and commerce were to be restricted to 65 per cent of their normal consumption of electricity. Continuous process industries were to be exempt from the three-day week but were ordered to reduce the level of production by a similar proportion. No legal restrictions were placed on domestic consumption but householders were asked to restrict the use of electricity for space heating to one room and then only if no other form of heating was available, and the BBC and the commercial companies were ordered to cease bradcasting at 10.30 p.m. each evening.

Shortly before Christmas, however, the centre of attention returned to the negotiations between the NCB and the NUM. On 20 December the NUM executive had one of its many meetings with government Ministers during the early part of the dispute, and it was encouraged by William Whitelaw, Secretary of State for Employment, to reopen negotiations with the NCB to discover if there was any possibility of improving the Board's offer without breaching Phase Three. When, the following day, the Board and the union met, the idea of payment for time spent at the pit before and after work was revived. Despite initially receiving an unencouraging reply from the Pay Board, the two sides agreed to press ahead and explore the idea further. For a while there was considerable optimism that this idea might provide a solution to the dispute. The NUM calculated that if the time spent before and after work at the pits was paid for at overtime rates it would yield general increases of a further £2-£3 a week for all miners. For their part the NCB agreed to pay the extra money if the Pay Board gave them permission to do so. In the event this attempt to solve the dispute came to nothing. In order to permit the increases under Phase Three the Pay Board required evidence that the time spent at the pit before and after work could reasonably be classified as coming within the official working day. The Coal Board calculated, however, that very little of the time under discussion could be classified in this way (payment for such time at overtime rates would only have amounted to an extra 70p a week.)

The collapse of the attempt to solve the dispute by payment for waiting time was the signal for the entry of the TUC into the negotiations. On 9 January a meeting was held of the TUC Economic Committee. It was presented with a paper prepared for a special conference of union Presidents and Secretaries which had been called for the following week. The paper argued that the settlement of the miners' strike in 1972 had not led to a series of inflationary wage settlements in other industries as had been suggested by the government. The basis of the argument was that for the four months following the Wilberforce award wage settlements continued at much the same level as they had done before. It was not until the summer of 1972 that wage settlements started to rise and this rise was attributable to changing economic conditions (rising prices, labour shortages and the like), not the influence of the Wilberforce award. Taking their cue from this document the TUC Economic Committee decided to press their previously expressed view that a wage rise outside Phase Three for the miners would not necessarily lead to similar wage rises in other industries. In fact, they went considerably further than they had ever done before by offering to conclude an agreement with the government to the effect that if the miners were allowed a wage rise outside Phase Three then this would not be used as an argument by other unions in their wage claims.

On the afternoon of 9 January the TUC leaders met the government and the CBI representatives at one of the regular National Economic Development Council Meetings. At this meeting Len Murray, TUC General Secretary, read out a statement outlining the proposed deal.

> The General Council accepts that there is a distinctive and exceptional situation in the mining industry. If the government are prepared to give an assurance that they will make possible a settlement between the miners and the National Coal Board, other unions will not use that as an argument in negotiations for their own settlement.[41]

The initial reaction of both the government and the CBI was hostile. Anthony Barber, Chancellor of the Exchequer, told the TUC that

> he could not accept the offer . . . the government could not agree to any settlement outside the limits of Phase Three . . . To accept the TUC suggestion would be unfair to the four million workers who had already settled within the terms of Phase Three.'[42]

Campbell Adamson, Director-General of the CBI, took a similar view: 'as far as we are concerned this is not a movement which could possibly lead to a settlement.'[43] Later, however, both the government and the CBI modified their attitude. The TUC were invited to discuss their plans with the government and for a while there was hope that it might be able to form the basis for a settlement. In the event, however, the plan was finally turned down because the government insisted that it needed 'guarantees' that member unions would adhere to the TUC's undertaking, 'guarantees' which the TUC felt it could never hope to provide. So despite further meetings between the government and the TUC and the ratification of the TUC initiative by a conference of Presidents and Secretaries of TUC-affiliated unions, the plan failed to provide a solution to the dispute.

At this juncture events took an interesting and in some ways crucial turn. The Opposition had for some time claimed that the government's decision to introduce the 'three-day week' had been ill-founded. The Opposition spokesman on Trade and Industry, Anthony Wedgwood Benn, claimed that coal supplies were much better than the government had led the country to believe. The government, on the other hand, had maintained that the three-day week was essential. If measures had not been taken to reduce the level of demand for electricity then the country would have been in a crisis situation. For example, the Leader of the House of Commons, James Prior, warned that if the government had not taken action the country might have faced 'a situation in which we could not get fresh water and we could have sewage floating in the streets of London and in other big cities.'[44] As late as the middle of January government Ministers were warning that not only was the three day week essential but that unless further economies were made even more drastic measures might have to be taken. Suggestions were made about legal curbs on the domestic consumption of electricity.

> In an urgent drive to cut still further the nation's consumption of electricity the government's new energy team, headed by Lord Carrington, is considering unprecedented legal controls over the use of electricity in the home.
> Such curbs on heating, lighting and the use of electrical appliances would be impossible to police comprehensively. But they could be subject to spot checks, and it is felt that most of the public would be readier to comply with legal restrictions than with the strongest request. Further measures to tighten up the energy squeeze could also mean less power for entertainment, such as cinemas and football matches, and food shops limited to half-day opening like all others.[45]

Within a matter of days, however, the mood changed. On 16 January, Arthur Hawkins, Chairman of the CEGB made a statement in which he said that he was pleased with the response of the public to requests for economy and he could see no reason why more restrictions should be needed. Further, on 17 January, Lord Carrington, Minister for Energy, announced that the position had substantially improved. Coal stocks at the power stations, he said, had declined by only 200,000 tons in the second week of January, compared to 600,000 tons in the first week and rates of one million tons around the end of November. Currently, he said, stocks stood at 13·7 million tons and if the rate of run-down were confined to that of recent weeks then it would be the end of August before they reached danger level. He concluded, therefore, that it might be possible to extend the working week. Talks were arranged with the TUC and the CBI for the following week to discuss such a possibility.

The change in the government's attitutde provoked an immediate response from the miners. Their negotiating team met and recommended that the NUM ask their members for permission to call a national strike.

Explaining the reasoning behind their decision, Joe Gormley said:

> This is because I have the feeling that the overtime ban is not
> having the effect we thought it would have in the early stages. This
> can be supported by the coal stock figures quoted by Lord
> Carrington last week. I feel that we have to become more positive
> if we are to succeed.[46]

The recommendation of the negotiating team was accepted by the
full executive (16 votes to 10) and voting in the strike ballot took place
on 31 January and 1 February. Miners were merely asked to say yes or
no to the following question.

> The national executive committee has rejected as unsatisfactory
> the board's offer to increase wages and fringe benefits (details of
> the offer have been widely circulated) and reocmmended the
> members to give the committee permission to call a national strike
> in support of our efforts to obtain a satisfactory response to our
> claim on behalf of all our members.
> Are you in favour of the national executive committee being
> given authority to call a national strike?[47]

Despite initial suggestions that the union might not receive the degree
of support it needed, the result of the ballot showed an overwhelming
majority in favour of strike action: 86 per cent of NUM members voted
and 81 per cent of these voted in favour of a stoppage. Further,
majorities in favour of strike action were obtained in all but one of
the areas (see Table 27) and some of the previously moderate areas
(such as Durham and the Midlands) voted heavily in favour of strike
action.

The decision of the NUM executive to call a strike from 10 February
effectively put paid to any possibility of a return to a four- or five-day
week. It did not, however, mark an end of all attempts to reach a
negotiated settlement of the dispute. The last major attempt to settle
the dispute before the commencement of the strike involved the
'relativities report' of the Pay Board. On 24 January the Pay Board
(which had been established to oversee the government's pay policy)
published the results of a study it had been making of ways in which
the pay code could be amended to enable exceptional awards to be
made to certain groups of workers.[48] The government's initial
reaction was to argue that the report would be of little assistance in
solving the miners' dispute. The Pay Board, the government argued,
had suggested a procedure for examining the relative position of
different groups of workers but such an examination would probably

take quite a while and was not anyway suitable for a 'crisis' situation. Later, however, the government appeared to change its position somewhat. In a television interview on 28 January the Prime Minister hinted that the relativities report might have some relevance for the miners' dispute. Three days later the *Daily Mirror* published an interview with the Prime Minister in which in reply to the question 'Can the relativities report help with the miners' dispute?' (a question which, it has been claimed, was inserted at the request of the Prime Minister) he said,

> I think it could. If the TUC and the CBI accept the Report, and we can get the procedure established quickly; if the miners resume normal working; then they can have the settlement which is available to them, and they can get their case fully examined.[49]

Early in February discussions began in earnest about the possibilities of using the relativities report to settle the strike. The unions had made their position clear. The miners were unwilling to return to normal working or to postpone their strike action unless there was an offer of 'more cash on the table'. The CBI, although they did not want to meet all of the miners' demands, were at least willing to meet them half-way. An emergency meeting of the CBI Council on 4 February passed a resolution approving the principle of the relativities report, agreeing that any award made to the miners under the relativities prcedure could be back dated and agreeing that if necessary Phase Three could be breached in order to reach a settlement with the miners. In their talks with the CBI and the TUC on 4 and 5 February, however, it became clear that the government was unwilling to go beyond merely saying that if the miners returned to normal working then their case could be referred to a relativities board. They would not agree to a settlement outside Phase Three and they appeared even to oppose the idea of the backdating of any relativities award (although it has since been suggested that they were not quite so intransigent on this point as was at first thought).

As a result, the talks came to nothing and the miners determined to go ahead with their strike. On 7 February the government announced that there was to be a General Election (on 28 February); they made it quite clear that the miners' dispute was one of the main reasons for calling the election and as far as they were concerned the dispute (and the philosophy surrounding it) was going to be one of the main issues in the campaign. Along with the announcement of the election the government appealed to the miners to postpone their strike action. Further, they proposed to ask the Pay Board to examine the relative position of miners' pay. In the event, the NUM executive decided to continue with the strike. They did, however, agree to give evidence to the Pay Board's inquiry.

The Pay Board began its inquiry on 18 February, in the midst of the election campaign. The NUM, in its evidence, centred its case on two points. First, they argued that the miners' average earnings had, since the Wilberforce Report, fallen considerably relative to those of other workers. In October 1973, they said, a miners' average weekly wage had been £37 compared with an average weekly wage of £42·2 in manufacturing industries. In April 1973 a miner had earned only 88 per cent of the average wage in manufacturing industries compared to 93 per cent after the Wilberforce Report, in April 1971. Second, they argued that wages in coal-mining were too low to attract men into the industry. Coal-mining was dirty and dangerous; it was an unattractive occupation to many men. Yet men had to be attracted to the industry if it was to fulfil the new demands likely to be placed upon it. Men would only be attracted if better wages were offered. In support of their argument that the present wage rates were insufficient to attract recruits they pointed to the evidence of recent manpower figures. In the last quarter of 1973 6,069 men had left the mines to find other jobs, compared to only 2,872, in the same period of 1972 and 3,384 in 1971. Again, in the third quarter of 1973 only 291 out of 1,583 vacancies in the industry had been filled.

The Coal Board, in its evidence, broadly supported the views expressed by the NUM. Presenting the NCB's case, Derek Ezra, Chairman of the Board, said,

> It is clear that the relative importance of coal-mining to the United Kingdom economy has increased and that it will continue to hold new importance in the future . . . Because of this it is the Board's view that corresponding improvements in the wages and conditions of service of the workforce are vital so that the challenge of this new role can be met successfully.[50]

The Board argued that in deciding what the level of miners' pay should be the inquiry should take four factors into account. First, recognition should be given to the psychological barrier to working underground. Second, allowance should be made for the arduous nature of the work. Third, note should be taken of the risks and health hazards involved in mining. Fourth, it should be recognised that mining in 1974 required 'special skills'. The Board did, however, qualify these remarks by saying that it would only support the claim for a substantial improvement in miners' earnings if the NUM agreed to the return of compulsory arbitration in the industry (compulsory arbitration had been abolished in 1961; the decisions of the National Reference Tribunal were no longer held to be binding) and the introduction of a local productivity bonus scheme.

The miners' case also received some if not quite such explicit support from the evidence of two government departments. The Department of Energy in its written evidence confirmed that the importance of the coal-mining industry for the future of the British Economy while the Department of Employment confirmed the physical dangers involved in mining. Even the Pay Board itself contributed a little to the miners' case. The Pay Board pointed out that the figures used in evidence had overestimated the average miners' pay in comparison with that of other workers. Although the intervention may not have been quite as important as was at first suggested, it clearly did nothing to detract from the miners' arguments. The only real problem raised during the hearings was raised by another group of workers. The railwaymen, in particular, highlighted the possibility that other groups might try to follow any 'exceptional' award given to the miners by claiming that workers in the railway industry were so clearly linked to the miners that they would expect the basis of any miners' settlement to be applied to their industry.

The Pay Board, in its report,[51] closely followed the spirit of the evidence it had received. The coal-mining industry was, it said, clearly a special case, and recognition ought to be given to that fact. Two points were stressed. First, it was essential to slow down the loss of manpower from the industry if it was to successfully meet the demands that were likely to be made on it in the future. Better wages were not the only way to slow down the loss of manpower but they were an important part of any attempt to do so.

> We do not believe that in this or any other industry pay and conditions are the only factors affecting people's attitudes towards remaining in it or being prepared to consider employment in it. In the case of coal we attach particular importance to the effect on these attitudes of the new prospects of a prosperous mining industry offering life-time careers. This is in sharp contrast to the background of contraction which has inevitably discouraged long-term commitment to the industry, especially among young trained mine-workers.
> But although all these conditions are important we do not believe that the coal industry will be able to recruit and retain sufficient mineworkers to sustain the current level of output of deep-mined coal — the required contribution to the nation's energy supplies — unless the pay of mineworkers is improved relative to that of other workers.[52]

Second, it argued that special consideration ought to be given to those who worked underground. They needed to be compensated for the uncomfortable and dangerous conditions in which they worked. Further if they were not paid more then it was doubted that the industry would be able to attract adequate numbers of recruits to perform crucial tasks underground.

In its recommendations, the Pay Board suggested that in addition to the Coal Board's original offer a special allowance of £1.20 a shift should be given to all those who worked underground. Surface workers who had previously worked underground should be eligible for a personal allowance of 60p. Further, the NUM and the NCB should be permitted to 're-cast' the original offer and with the help of an extra £4 million should be able to make necessary adjustments to internal relativities. The Pay Board estimated the cost of their award at £57·5 million.

The Pay Board report did not in the event play quite such a crucial role as it was at one time thought it might. The election brought a minority Labour Government to power and the new Secretary of Employment, Michael Foot, permitted the NUM and the NCB to renegotiate freely without reference to the Board's report. The final settlement was estimated to cost £103 million (about the same as the Pay Board's recommendations, taking the Coal Board's original offer into account). New minimum wage rates of £31 a week for surface workers, £36 a week for underground men and £45 a week for faceworkers were agreed. In addition, the Board agreed to pay a 19p an hour premium for 'anti-social' hours worked at night, increased holiday pay, an increased lump sum to those who voluntarily retired from the industry (£500 instead of £300) and improved death-in-service benefits. The new package was approved by an overwhelming majority of the NUM national executive committee and by miners' representatives throughout the coal-fields.

The 1973-4 miners' dispute was probably the most important industrial dispute in Britain since the 1926 General Strike. In the course of the dispute it became clear that it was not a purely domestic dispute between the NUM and the NCB; it was really a conflict between the government and the miners' union. There is little doubt, of course, that the government took a keen interest in and exerted a considerable influence on the course of the 1971-2 dispute. During the 1973-4 dispute, however, the government's role was much more transparent than it had been two years earlier. In particular, the government's role was more clearly defined in the later dispute because of the existence of pay legislation which virtually prohibited a settlement on the terms the miners were demanding (and, from the information available, it would appear the terms on which the NCB was willing to settle). The only agent who could change the pay code or had the necessary authority to bypass it (under certain circumstances the pay legislation permitted the government to bypass the rules laid down) was the government.

It is also probably true that to an extent the dispute was in many ways not only between the miners and the government but also between the trade union movement as a whole and the government. While the trade union movement was not initially directly involved in the dispute (it played a more active role in the later stages) it clearly took a keen and partisan interest in it. The miners' dispute was one which the trade union movement could not afford to lose, largely because of its symbolic importance. There is little doubt that the trade union movement, which had not enjoyed a great deal of success in its disputes with the government over the Industrial Relations Act, was determined that it should not fare so badly in the dispute over the government's pay policy. Defeat for the miners would not necessarily have led to defeat for the rest of the trade union movement in this arena, but it would have gone a long way towards doing so.

In the end the dispute became almost centrally a political event. The General Election may not have been caused by the miners' strike, but it was certainly an important precipitating factor, and the Conservative Party were determined that the strike and the philosophy behind it should be a major platform in the election. The issue of 'Who governs Britain?' was not merely related to the miners' strike – it took in at least four years of conflict between the government and the trade unions – and it was not anyway the only issue in the election, but it was clearly a major one.

By most measures the miners must be judged to have been successful in the 1973-4 dispute. They did not, of course, gain everything they had claimed in their original demand; but very few trade unions expect to do so. One of the curious features of the strike was the feeling, often expressed in the news media, that the central question was not whether the miners would 'win', but when and how they would 'win'. It seemed as though almost everyone was trying to find a way to give the miners most of what they were demanding but could not do so without breaching the Phase Three legislation. To a certain extent, this is an oversimplification, for on occasions the government seemed determined to take a stand 'against' the miners, yet even this impression may be misleading, for what is probably nearer the truth is that the majority of the government were keen to take a stand 'in favour' of the pay policy, but would have been happy if the miners could have been satisfied without explicitly breaching this policy. Certainly if the suggestions which were made that the pay policy was specially designed to enable the miners to obtain a satisfactory settlement are correct, then this would seem to support the argument.

160

As with the 1971-2 dispute, there were undoubtedly a host of factors that influenced the outcome of the 1973-4 dispute. It might be worthwhile, however, mentioning some of the important ones.

First, it is clear that although the miners did not have the degree of public sympathy they enjoyed during the 1971-2 dispute they nevertheless retained a considerable amount of it. The government in 1973-4 seemed to have learned from the lessons of their earlier experience and made strenuous efforts to persuade the general public that a 'stand' against the miners was essential in the 'national interest'. Yet despite the restrictions (some of which, it has been suggested − such as the broadcasting restrictions − at least to a certain extent were introduced in an attempt to emphasise the importance of the dispute to the general public) and despite discussion about the influence of communists on the miners' executive, the miners retained a good deal of public support, certainly far more than many other unions have enjoyed during disputes. The attitude of a considerable section of the public was contradictory but helpful to the miners. It was probably best illustrated by the results of a public opinion poll conducted during the dispute and reported in the *Sunday Times*, which showed that a majority of people interviewed supported the pay code but also supported the idea of giving the miners most of what they had asked for (even if this meant breaching the pay code). This attitude is probably best summed up in the feeling that the miners did an unpleasant job which was not adequately rewarded and that although wage settlements ought to be restrained in order to control inflation, the miners were the wrong target, and anyway a fight with the miners 'just wasn't worth it'.

Second, during the 1971-2 dispute the miners had not always been sure of the whole-hearted backing of the rest of the trade union movement. The TUC, for example, had declined to arrange the round table conference of other unions requested by the miners. In 1973-4 the rest of the trade union movement was much more firmly committed to supporting the miners. The TUC itself was willing to go to considerable lengths to try to persuade the government to treat the miners as a 'special case'.

Third, the miners' hand was obviously strengthened by the power crisis resulting from the oil restrictions. This helped in two ways. First, it meant that they possessed a much better bargaining weapon than they had held for the past fifteen years or so. Few people argued in 1973-4 as they had done in 1971-2 that the country could well manage without a coal-mining industry. Second, it meant that their argument about the need to raise miners' wages in order to attract men into the industry was widely accepted. Even the government accepted the force of this argument; their position was not to question whether miners' wages needed to be raised, but when and how.

Fourth, and in some ways most crucially, the miners were helped by their success in the 1971-2 dispute. The resolve of the miners in 1973-4 was clearly strengthened by their belief that they would succeed in achieving their objectives even if it took a while to do so. Two years earlier few commentators, and possibly even few miners, believed that they could succeed in such a dispute. The 1971-2 dispute had shown that the miners held greater bargaining strength than most people had realised. In the far more advantageous circumstances of 1973-4 it seemed almost inevitable that this bargaining strength would bring them most of what they were demanding.

The Future

In early 1974 the future of the coal-mining industry looked far better than it had at any time since the decline of the markets in 1957. There was widespread recognition of the vital importance of coal for the future of the nations' economy. Coal was a 'secure' source of power and, at least until the North Sea oil potential is realised in the late 1970s, Britain's major indigenous source of power. Further, the incredible rise in the price of oil had resulted in coal being competitive, in its own right, at least in terms of price. The Pay Board in its report on miners' earnings suggested (on the basis of evidence presented by the Department of Energy) that in the future, provided 'world oil prices remain at about their present level and the costs of (coal) production do not rise to an exaggerated extent, the NCB will be able to sell as much coal as it can provide.'[53]

Nevertheless, despite these encouraging portents there were still those who were willing to argue that coal's future might not be quite as secure as many imagined. *The Economist* in early 1974 ran a series of articles in which it suggested that the present energy shortage might be replaced in a few years by an energy 'glut', because there would be an over-reaction to the shortage on the part of governments and private companies. Further, the coal-mining industry might in the near future find its price advantage eroded. Central to this idea was the notion that the current high price of oil would be reduced in a short while. In particular, it was believed that Saudi Arabia, because of its large oil reserves, would want oil prices reduced in order to ensure that the trend to find substitutes for oil was not carried too far, and that if they could not obtain the agreement of other oil producers in this move, then they could take unilateral action by 'flooding the market'.

Some support for this latter point of view came from the oil industry itself. Many companies believed that the oil producers had 'over-reached' themselves with the size of the increase in the price of oil and that they would consequently be forced to reduce prices in the future. Even the Pay Board, who had helped to paint the optimistic picture for the future of the coal-mining industry recognised the dangers.

In round terms, the price of power station fuel oil might be about 6p a therm, and the average cost of power station coal, taking account of the current NCB offer, something less than 5p a therm. But a reduction of around 10 per cent in the world price of oil might close that cost advantage for coal by as much as 0·5 per therm. It has also to be remembered that because some coal burning power stations are old and inefficient, the attractiveness of coal as a power station fuel for the CEGB gets smaller as the amount of coal used increases.[54]

It is of course difficult to predict future events in an area such as this. It is just as likely that in the near and medium-term future oil prices will rise, as it is they will fall. If, however, they do fall and oil becomes more easily obtainable then once again the future of the coal-mining industry might look less than secure. The government, of course, in such an eventuality, could take measures to protect the coal-mining industry; possibly more effective measures than were taken in the 1960s. It may be that the experience of recent evens will have convinced the government and future governments that even if there is an alternative cheaper source of energy, the coal-mining industry is still worth supporting. Certainly it is likely that if the coal-mining industry suffered yet another reversal in its fortunes it would not at some later date be able to respond to a revival in demand. It has become clear during the last few years that the industry suffered far more during the 1960s than was indicated merely by the size of the contraction of the industry. The decline persuaded many, particularly younger workers, that there was no future in the industry and there was little point in looking to it for a career. Unless the industry can attract a larger number of young recruits, and it is doubtful if it could hope to do so if it suffered yet another decline, then there is little hope for the future.

NOTES

1. The break in the pipeline meant that Libyan oil was a much more attractive product than many of its competitors because of the relatively short journey to European markets.
2. *Coal News*, February 1971, p.1.
3. Op. cit.
4. *Coal News*, August 1971, p.6.

5. *Coal News,* October 1971. p.6.
6. *Coal News,* December 1971, p.8.
7. *Coal News,* December 1971, p.8.
8. *The Economist,* 11 December 1971, p.85.
9. *The Economist,* 4 December 1971, pp.74-7.
10. *The Times,* 11 January 1972, p.1.
11. *The Times,* 10 January 1972, p.1.
12. *The Times,* 10 January 1972, p.1.
13. *Guardian,* 12 February 1972, p.1.
14. *Guardian,* 15 February 1972, p.5.
15. *Sunday Times,* 13 February 1972, p.12.
16. *Guardian,* 14 February 1972, p.18.
17. J.Hughes and R.Moore (eds.), op.cit.
18. Op.cit.
19. Ibid., para.20.
20. Ibid., para.32.
21. Ibid., para.34.
22. Ibid., para.35.
23. They gained ten further concessions:
    (i)   Personal and piece rates to be adjusted;
    (ii)   Similar increases for coke and by-product workers;
    (iii)   Canteen and other workers to have increases negotiated
            backdated to 1 November;
    (iv)   Clerical workers to have a £5 a week increase;
    (v)   Juveniles to get adult rates at 18 in two years time;
    (vi)   Winding enginemen upgraded and paid an extra £5.80;
    (vii)   New five grade structure for lorry drivers from £23 to £24.50
            a week;
    (viii) Extra five days holiday from 1 May;
    (ix)   Immediate discussions on productivity deal leading to pay rises
            by September;
    (x)   Miners' transport to be subsidised on a national basis.
24. *Sunday Times,* 20 February 1972, p.12.
25. *Sunday Times,* 13 February 1972, p.45.
26. *Scottish Daily Express,* 19 February 1972, p.8.
27. *Guardian,* 15 February 1972, p.12.
28. C. Kerr and A. Siegel, 'The Inter-Industry Propensity to Strike – An
    International Comparison', in A. Flanders (ed.) *Collective Bargaining,*
    Penguin, Harmondsworth, 1969. See also G.V. Rimlinger, 'National
    Differences in the Strike Propensity of Coal Miners: Experience in Four
    Countries', *Industrial and Labour Relations Review,* Vol.12, 1959, No.3.
    pp.389-406, for a note of reservation.
29. *Guardian,* 12 February 1972, p.12.
30. *Guardian,* 7 February 1972, p.12.
31. *Guardian,* 12 February 1972, p.12.
32. *Coal News,* July 1972, p.11.
33. Coal Industry Act, 1973.
34. *Petroleum Press Service,* November 1973, p.402.
35. A series of measures designed to counter inflation by controlling wage
    rises and price increases. Phase One introduced a wages and prices standstill;
    Phase two permitted certain small wage rises (based on a norm of £1 plus
    4 per cent).
36. Reported in *Sunday Times,* 3 February 1974, p.12.

37. *The Times*, 11 October 1973, p.1.
38. *Sunday Times*, 3 February 1974, p.12.
39. *The Times*, 9 November 1973, p.1.
40. *The Times*, 7 December 1973, p.1.
41. *The Times*, 10 January 1974, p.17.
42. *The Times*, 10 January 1974, p.17.
43. *The Times*, 10 January 1974, p.17.
44. *The Times*, 31 December 1973, p.1.
45. *Sunday Times*, 13 January 1974, p.45.
46. *The Times*, 23 January 1974, p.23.
47. *The Times*, 26 January 1974, p.1.
48. Pay Board, *Problems of Pay Relativities*, Cmnd. 5535, HMSO, London,1974.
49. Quoted *Sunday Times*, 3 February 1974, p.13.
50. *The Times*, 20 February 1974, p.3.
51. Pay Board, *Special Report : Relative Pay of Mineworkers*, Cmnd. 5567, HMSO, London, 1974.
52. Ibid., p.22.
53. Ibid., p.6.
54. Ibid., p.5.

7. CONCLUSION

Throughout this book the aim has been to examine changes in the 'market position' of the coal-mining industry, the consequences of these changes and the reaction of various parties to them. In this concluding chapter an attempt will be made to draw together some of the threads of earlier discussions, and to examine more centrally the reactions of governments, employers and miners (plus their unions).

Governments

Governments (and one is talking, of course, about a variety of governments of different political complexions) during the early part of the period under discussion intervened on a number of occasions in the affairs of the coal-mining industry. Frequently, though, even if they knew what they wanted to do they rarely had sufficient fervour to carry their ideas to a successful conclusion. During both the First and the Second World Wars governments recognised the importance of maintaining coal supplies. Without adequate supplies of coal the war effort would have been seriously hampered; at one time during the Second World War Bevin, Minister of Labour, commented that the shortage of coal was one of the most serious problems facing the government in the war effort.[1] Yet, despite this recognition of the importance of the role the industry had to play, the government rarely carried through measures effective enough to deal with the problems. Initially in each war they hoped that the industry itself would carry through the necessary reforms, when they really had little reason, especially during the Second World War (after the experience of the previous decade) to expect that they would do so. The comment of *The Economist*[2] about the performance of the government during the Second World War — that they could hardly have done less than they did and that they retreated at the first growl from either side — fairly accurately sums up government activity during this period. The dual control that was finally adopted during both wars, after voluntary methods failed, was really the result of the government's spirit of compromise — it satisfied nobody, but at the same time it did not outrage anyone either.

Between the two world wars government policy continued in much the same vein. Immediately after the First World War it appeared as if the government did not really know what action to take. It certainly did not put forward a clear policy but rather seemed content to provide a forum for the two sides of the industry to fight out their own battles. The way that the Sankey Commission[3] was set up and events leading to the General Strike are good examples of this kind of attitude. An uncertain *'laissez faire'*, 'hope the problem will go away' attitude prevailed.

After the General Strike governments seemed to have a clearer idea about what ought to be done with the industry. Changes were essential to rationalise the industry and improve its overall structure. Yet although governments seemed to understand the kind of action that needed to be taken they seemed unwilling to press their views on a recalcitrant industry. The 1926, 1930 and 1938 Acts[4] were all attempts to persuade the industry to move in this direction but none contained the powers necessary to ensure that the essential changes were made.

After the Second World War the situation altered. In 1945 a Labour Government was elected to power with a platform that declared in favour of nationalising the industry. The government moved with considerable speed introducing the Bill and carrying it through to a successful conclusion in the space of a year. By 1 January 1947 the industry had been taken into public ownership.

The government's role was henceforth to be rather different. No longer were they to sit on the sidelines, prodding, criticising, encouraging the industry to make the kind of moves they felt were necessary. They were now in a position to influence the industry's fortunes in a much more direct fashion. Although the structure adopted gave day-to-day control of the industry to an independent body, the National Coal Board, the government had an important role to play. It appointed the Board, had the power to give general policy directives, and exercised a crucial influence over the industry's finances. The power the government enjoyed to control the industry's borrowing, and through a private agreement the price of coal, was important in this latter context.

The result was that after nationalisation it was the government, not the Coal Board, that decided a great deal of general policy. The Coal Board could advise and warn, but in most cases it could not enforce its views in this area.

There is little evidence of major differences between the Coal Board and the government over general policy in the immediate post-war years. Both saw the need for the industry to expand output as quickly as possible in response to rising demand. One of the few important areas of disagreement concerned the price of coal. The Coal Board were prevented by the government from raising the price of coal to something nearer its market value (and in the event deprived of a great deal of revenue which might have helped to ease the Board's financial problems in later years).

The views of the Coal Board and governments began to differ more significantly after the decline of the markets. At first the government, like the Coal Board, felt that the fall in consumption was probably a temporary phenomenon and would soon be reversed. When it became clear that this was not the case then governments initially encouraged, but later put considerable pressure on, the NCB to reduce the size of the industry. From 1965 onwards government forecasts were always of a future demand for coal of a smaller order than that suggested by the NCB. On most occasions they emphasised that their figures were forecasts not instructions, yet through the considerable powers they held they ensured that their forecasts had an important impact on Coal Board policy. Governments, of course, were keen to cushion some of the worst effects of the run-down of the industry and spent considerable sums doing just that. At times they were also willing to take measures to slow down the decline of the industry (as in 1967) so that such events would not have too catastrophic a social effect. Yet these moves to slow down the industry's decline were not attempts to reverse market trends. They were a recognition of the appalling social consequences of a rapid decline, but essentially aimed to lengthen the period of the decline rather than to stop it altogether.

Since the early 1970s government policy has changed somewhat. It has started moving towards the view that the coal-mining industry may have a more significant role to play in the economy of the nation. The control of oil supplies and the increase in oil prices have crucially affected this attitude. This is not to say that the government necessarily envisages a tremendous expansion in the size of the industry but it now seems to believe that the decline should be halted. In this, of course, they have the support of the Coal Board. However, there have still been areas of friction. In particular, the government in 1971-2 and 1973-4 seemed to believe that its general pay policy was more important than the specific needs of the coal industry. To a large extent it seems that during this period the Coal Board was much keener than the government to meet the demands of the miners' union.

Recently governments have been criticised for their previous fuel policies. Clearly on many occasions their forecasts have proved to be incorrect. This in itself should cause little surprise. Forecasting and planning in general are very imprecise exercises and few people have claimed that fuel forecasts can be made with any greater accuracy than can forecasts of any other social or economic event. The policies of various governments towards the coal-mining industry since the Second World War, however, might justifiably be criticised on two counts.

First, although most people recognise that planning and forecasting are impreciase exercises, and although the government itself has

168

frequently stressed this in relation to fuel policies, governments have not always acted as though this were the case. Frequently they have been slow to react to changes which have radically altered the assumptions on which their plans and forecasts have been based. It has been almost as though they have retained faith in their plans until it has become quite obvious that they were wrong. Maybe one of the reasons for this is the political difficulty involved in recognising that one has made a mistake. Whatever the reasons, both in 1959 and 1973 governments do not seem to have reacted to warning signs as quickly as they might have done. On each occasion warning signs were visible about two years before effective action was taken. Many observers warned that the pattern of demand for coal was changing in early 1958, and the vulnerability of oil to the kind of action that was taken in 1973 had been illustrated between mid-1970 and mid-1971 when oil prices were almost doubled following the break in the Trans Arabian pipeline. If one recognises the imprecise nature of forecasting and planning then deviations from the expected path ought to be treated as warning signs and the plan ought to be reconsidered. In fact governments do not seem to have acted in this way. They seem to have assumed that deviations, because they did not fit in with the forecast, must have been temporary and have only reconsidered the plan when it has clearly had no basis in reality.

Second, governments might be criticised for letting themselves be led too firmly by the market. There is surely a case for arguing that fuel and power are so vital to the well-being of a nation that a government should be over-cautious. Is there, for example, any real reason why a government should not plan for a surplus of power, especially if certain of its sources of power are not entirely secure? To link this to the particular case of the coal-mining industry, should governments during the 1960s have been so willing to give in to market forces as they were? Was there not a case for arguing that the coal-mining industry should not have been 'run down' to the extent that it was, even though its product may not have been as attractive as some of the possible substitutes? As the government found in 1973, once an industry has been 'run down' to the extent that occurred with coal-mining it is not easy to reverse the process. It takes a while to recruit and train miners, and probably longer to reconstruct old or open up new mines. Such a policy would necessarily have involved either further government subsidies for the coal-mining industry or further control over substitutes, but given the crucial role of fuel and power are either of these too high a price to pay?

Employers

The coal-owners, who controlled the destinies of Britain's coal-mining industry until 1947, have been described as 'some of the fiercest individualists anywhere in the world'.[5] Evidence to support this statement is not hard to come by.

Almost all coal-owners opposed government intervention in the industry's affairs. Not only did they oppose, as one might expect, the moves to nationalise the industry, but they also opposed the limited form of government intervention adopted during both world wards and government attempts to promote amalgamations. The manner of their argument against the most important move made by governments in this latter direction, the 1938 Act,[6] is instructive, for they argued that the Act took away their traditional safeguards. As *The Economist* pointed out,[7] these 'traditional safeguards' meant a barrier to any successful government intervention in the industry's affairs.

Many, although not all, coal-owners went even further than this. Not only did they oppose government intervention to reorganise the industry but they also opposed moves by the industry to reorganise its own structure. Reorganisation would have meant that many of the multitude of small coal-owners would have lost their independence and they were unwilling to accept such a possibility. They argued that such moves were unnecessary. The industry's problems could not be solved by amalgamations and the like; the industry's problems centred around its cost structure and its competitive position. It was this kind of attitude that led to the moves to reduce wages in the 1920s and the General Strike of 1926. It was only, in fact, when nationalisation appeared to be the alternative that they made a 'death bed' repentence through the Foot Plan[8] and even that repentance was only partial, for the plan included numerous references to the importance of maintaining the freedom of independent action.

When the industry passed into public ownership in 1947 it was placed under the direction of the National Coal Board. The Board faced a number of problems in its first few years. The coal shortage of the 1946-7 winter was the most spectacular but only one of many. Clearly the shortage of 1947 could not to any marked extent be 'blamed' on the Coal Board; the origins of the shortage went back not only beyond the Board's assumption of powers, but also beyond the introudction and passage of the nationalisation Bill.

The extent to which the Board 'succeeded' in overcoming its problems was debated throughout the late 1940s and early 1950s. There seems little point in attempting such an assessment here, if only because the end product is likely to result in more questions than answers. Yet in one respect the Board clearly did succeed. This was in the area of cooperation with the industry's main trade union, the NUM. Throughout the late 1940s, the 1950s and the 1960s, the Board received as much cooperation in carrying out its tasks as any employer, even public employer, can ever expect in Britain. The NUM supported the Board not only in the period of expansion up till 1957, but also in the much more difficult period after 1959. To a certain extent the cooperation was a result of Union support which would have been forthcoming without any action on the part of the Board (it was the result of the union's long standing support for nationalisation). Yet the degree of cooperation was also to an extent the result of positive attitudes and actions by the Board towards the union. The fact that many ex-union officials were members of the National and Divisional Boards clearly helped, but so did the way the Board tried to consider and consult national union officials about Board policy. Not surprisingly such a policy earned the Board a great deal of criticism from certain quarters, particularly from production staff.

After 1959 the Board made considerable efforts to improve the industry's competitive position. Coal had to be produced as cheaply as possible; this meant mechanisation and concentration of production. Output per manshift rose impressively during the 1960s. Coal, however, did not only have to be produced but also had to be sold competitively. This meant that for the first time the Board had to 'sell' coal rather than merely 'share it out'. Thus, moves were made to improve coal's 'image', strengthen the marketing side of the Board's operations and reorganise distribution facilities.

It is also probably true that although they recognised that the concentration of production was essential if the industry was to be competitive the Board fought as hard as they could to halt the industry's sudden decline after 1959 (this was undoubtedly one of the reasons why they were able to retain so much union support during this period). The Board, and the chairman for the greater part of the period, Lord Robens, waged an almost ceaseless campaign for an increase in the production targets for the industry. An increase in the production targets would not, because of the need to concentrate production, have meant a reversal of the moves to reduce the size of the industry's labour force, but it would have meant a slowing down of this process. The Coal Board, however, was not 'the master in its own house', and the industry, despite the Board's protests, was considerably reduced in size and importance in the 1960s.

It is, of course, speculative, but it would be possible to argue that the coal-mining industry might not have been run down to the extent that it was after 1959 if it had been under private rather than public control. The reasoning behind this view does not point to the virtue or 'greater social conscience' of private industry. The reasoning rather centres on

171

the reactions of the miners and their unions. It is difficult to believe that if the industry had been under private control the miners and their unions would have been quite so cooperative in the 1960s. It is quite probable that had private employers tried to run down the industry to the extent that the Coal Board did after 1959, both the miners and their unions would have fought much more strongly against such action and although it is unlikely that they could have stopped any kind of decline at all they may well have succeeded in reducing the rate of decline.

Miners and Their Unions

Two features dominated the policy and actions of the miners and their unions during this period. The first concerns their desire to see the industry nationalised, and once it had been nationalised to ensure that nationalisation worked and was seen to work; the second concerns the essentially 'market' and 'economic' attitudes which ran through their reactions to events in the industry.

The MFGB declared its support for the nationalisation of the industry before the beginning of the twentieth centurey and although it is clear that certain miners and their leaders opposed such a policy to begin with, it is doubtful if this latter view held much sway by the outbreak of war in 1914. The clearest exposition of the attitude of the miners' unions towards nationalisation can be seen in the evidence presented to the Sankey Commission[9] in 1919. The miners' unions (and from the evidence available, almost certainly the majority of the miners) wanted the nationalisation of the industry with a considerable degree of 'workers' control'. One ought to be a little cautious, however, about the use of this term 'workers' control', for although it has been stated that the miners' evidence to the Sankey Commission was the 'clearest exposition' of their attitude towards nationalisation, it was still not entirely clear on this matter of 'workers' control'. It seems, in fact, as if the form of control proposed for the industry was interpreted differently by different miners' leaders. Some saw it as a move towards 'Guild Socialism' while others saw it merely as a way of giving the miners a 'say' in the running of the industry through a system of 'joint control'.

Whatever interpretation is put on the evidence of the miners' unions to the Sankey Commission, the nationalisation movement clearly suffered a set-back in the years following the publication of the Commission's Report. The refusal of the government to accept the majority report's recommendations marked the end of the first important 'drive' to achieve the nationalisation of the industry. For the next fifteen years nationalisation took second place in the miners' minds

172

and the unions' demands.

Nevertheless, nationalisaton was always part of the vision of the bulk of the miners and their unions, and the claims of nationalisation began to come to the fore once more during the early 1940s. The experiences of the war persuaded many previously uncommitted observers that nationalisation was essential for the future well-being of the industry and when the Labour Government gained office in 1945 moves in this direction seemed inevitable.

The attitude of the miners, their unions, the TUC, the Labour Party and socialist opinion in general over the form of control that should accompany nationalisation, however, had changed in the intervening years. Neither 'workers'' nor 'joint' 'control' was the aim of the bulk of the influential leaders. The TUC, the miners' unions and the Labour Party had all declared their opposition to such systems and rather favoured control by an independent state-appointed body.

It is also clear, however, that when the industry was nationalised in 1947 many rank and file miners expected bigger changes in the day-to-day running of the industry than actually occurred. By the end of the 1940s researchers were reporting strong support for the notion of greater 'workers' control' amongst a fair section of working miners. It is by no means certain, though, what interpretation one should put on this matter. It is not clear, for example, whether one should interpret this as meaning that individual miners had retained their faith in 'workers' control' while the Labour Party and the unions were changing their minds in the 1930s? Similarly it is not clear whether the expression of support for workers' control' in the late 1940s should be interpreted in a pragmatic or political light? The evidence that one has available, and it is by no means conclusive on this matter, leads one to believe that the majority of miners did move against the notion of workers' control in the 1930s and that their support in the late 1940s was more of a reaction against the conditions they found than anything else; it was, if you like, a negative rather than a positive reaction.

Nevertheless, while this is probably true of the bulk of miners one must be careful about generalising too much. It seems likely that many miners retained a greater degree of support for the notion of 'workers' control' than has so far been suggested, in particular, in certain parts of the country. In Scotland, for example, one can find far more evidence of support amongst miners (and for that matter amongst the work force in general) for the notion of 'workers' control' than elsewhere. Miners' leaders in Scotland fairly consistently opposed, and still do oppose, the majority of their English counterparts over this matter.

The support of the miners' unions for nationalisation extended well beyond 1947. Not only were they keen to see the industry nationalised but they were also keen to ensure that nationalisation worked and was seen to work. Mention has already been made of the degree of support the NUM gave to the NCB during the post-Second World War period. This support has been maintained to a large degree throughout the early 1970s. The NUM leaders made it clear in both the 1971-2 and 1973-4 disputes that their main quarrel was with the government, not the NCB.

One possible consequence of the NUM's support for the NCB's policies is that it may have led to a degree of disenchantment between the union, or certainly its national officials, and the rank and file membership. Unofficial strikes are by no means concrete 'proof' of a disagreement between the national leadership and the rank and file members of a union. There is a good deal of evidence that unofficial strikes can be the result of one of a number of other factors; the national leaders may just not have heard about them in time, the union may want to avoid giving strike pay, and so on. Many of the unofficial strikes in the coal-mining industry since the Second World War, however, do show signs of being caused, at least in part, by disenchantment between national officials of the NUM and rank and file miners. Certainly on a number of occasions the national leaders have not approved of the strikes and they frequently made attempts to persuade the miners to abandon their action. Further, many miners have argued that although the NCB may have consulted national union officials about policy decisions, they have not adequately consulted rank and file miners (there is still a considerable amount of criticism about the operation of the industry's consultative machinery).

Again, though, one should be careful about generalising too much. The degree of rank and file disenchantment with national union policies has by no means been uniform throughout the country. Certain areas, such as Scotland, Wales, and recently Yorkshire, have shown more evidence of dissatisfaction than others. It might also be noted that the disenchantment of rank and file members with national union policies has not led to the emergence of an established unofficial movement, in opposition to the union, as it has in some other industries.[10] Unofficial action in coal-mining has tended to be concentrated in certain areas and on certain issues; when an unofficial strike committee has been set up then it has usually been disbanded once the strike has ended. There are a number of reasons why this has happened. First, the NUM has a far greater hold in mining than many unions have in other industries. The NUM lodge in many cases forms a focal point for the local community. Second, the miners have over the years learnt the importance of collective action and a strong trade union; they have also learnt the dangers of rival sources of allegiance (the classic case, of course, is that of the 'Spencer Unions'). Third, much unofficial action has found support in the official union structure. The decentralised structure of the NUM, and the tradition of independent action of the various areas of the union, has meant that local union leaders have often been sympathetic towards and willing to support unofficial action even if this has meant opposing the official national union leaders.

It is also possible that in future the disenchantment between national union leaders and rank and file members might be reduced. The strikes of 1972 and 1974 were the first official national strikes in the coal-mining industry since 1926. To some miners this may indicate that the union is willing to press their case more forcefully than in the past. It is possible, of course, that a more forceful approach by the NUM may result in more official but fewer unofficial strikes, although this reasoning should not be taken too far as undoubtedly many of the various 'other causes' of unofficial strikes will remain.

The second feature of the policy and actions of the miners and their unions during this period concerns their 'market' and 'economic' attitude towards events. Put simply and briefly, the reference to a 'market' attitude is meant to indicate an acceptance by the miners that their fortunes would be closely tied to the 'market conditions' affecting the industry. To a certain extent, of course, all workers have to accept this; they accept that their ability to press wage claims and the like will be affected by the profitability of the industry and thus the market position of the industry in which they work. Similarly, in common with most other workers, miners have never merely accepted market conditions; they have always fought against wage reductions, the contraction of the industry and the like. The miners do, however, seem to have accepted, especially before the Second World War, a much closer relationship between their position and the market position in which the industry found itself; certainly they seem to have accepted that changes in wage rates, for example, would result from and be more closely tied to, much smaller movements in the industry's market position than has been the case with other workers. Undoubtedly the best example of this can be seen with the wage agreements of the 1920s, in particular the 'Terms of Settlement'. This tied the miners' wages not only to the national market position but also, and probably more crucially, to the market position of the individual coalfields. The owners managed to retain district wage assessments right up to the Second World War.

Since the war conditions in the industry have changed and national has replaced district bargaining. It may well be, however, that many miners and their leaders have retained their pre-war attitude on the importance of market conditions. It would hardly be surprising if this were the case. The pre-war years had a tremendous impact on most working miners and the social isolation of many mining communities has meant that past experiences have not been quickly forgotten. Past events are part of the fabric of many pit communities and are almost as much a part of the background of miners who never experienced them as they are of miners who were working in the industry at the time. Just, then, as miners found it difficult to forget the bitter struggles with the employers of the pre-war years, so many miners might have found it difficult to forget the effects of the fluctuating fortunes of the industry in the pre-war years.

This may, therefore, help to explain why many miners have found it difficult to understand the charges that were made against them in the early 1970s about 'holding the country to ransom'. During the early 1970s the miners used the improved market position of the industry to retrieve some of the ground lost during the 'depression' of the 1960s. It may well be that in the future further action will be taken, supported by similar reasoning. Why, they ask, are charges made that they are 'holding the country to ransom' when all they are doing is to respond, in the same way as they have so often done, and have been expected to do, in the past?Few people complained about the way market conditions influenced miners' wages in the 1960s; why should they suddenly complain about the way they influence them in the 1970s?

The reference to the 'economic' attitude of the miner and his unions is meant to indicate that by and large their concern, throughout this century, has been with economic matters. They have essentially sought to improve the miners' economic position within the present capitalist system, rather than to overthrow the system altogether; they have sought to use the system to their best advantage rather than to destroy it. Despite some of the comments of politicians and the news media to the contrary,[11] such a statement seems to hold for all of the industry's major disputes, including those in 1971-72 and 1973-74. This, in many ways, fits in with the miners' political background. Although a number of early labour leaders emerged from the ranks of the miners, as a group the miners were fairly slow to move towards socialism. They were the last major union to leave the Liberal Party and even when they formally made this break many retained a degree of sympathy for that brand of politics. Today the miners are amongst the stalwarts of the Labour Party but as a whole the miners' leaders are nearer to the centre than the 'left wing' of the Labour Party. There is, of course, today, as there has always been, a significant representation of the 'left wing' of the Labour Party and Communist Party on the national executive of the NUM but this representation, although it may be stronger today than it has been in recent years, is still, as it has generally been in the past, outnumbered by the contrary view.

176

It might be questioned whether this statement about the miners' economic attitude would cover the policy of the miners and their unions towards nationalisation. To a certain extent, at least, it would, because a considerable body of the support from the miners and their unions for the nationalisation of the mines had a 'non-political' basis. Even the Scottish leaders of the late nineteenth and early twentieth centuries stressed the benefits nationalisation would bring in terms of improved safety, welfare, earnings and stability; little stress was placed on the broader political effects of the action.[12] This kind of attitude was also evident in the 1940s when the miners' leaders argued that nationalisation would result in a better organised and administered industry; they stressed that they did not want to take any direct part in the administration of the industry themselves. The only evidence which counters such a point of view can be seen in the fairly spasmodic support for 'workers' control'. In the immediate pre- and post-First World War periods, for example, the idea of workers' control gained considerable support. It is doubtful, however, if even during these periods, which were in many ways the high point of the 'workers' control' movement in Britain, a majority of miners accepted the full impact of such ideas. Many it seems would only have moved part of the way along the road to 'workers' control' — an attempt to give the workers a greater 'say' in the way the industry was run — and few supported 'workers' control' for clear political motives. This is not to deny, of course, that there was a body of opinion within the mining community which supported the notion of 'full workers' control' and did so for 'political' reasons; there is not a great deal of evidence, however, to suggest that they were ever in the majority.

NOTES

1. See *The Times,* 21 July 1943, p.2.
2. 11 April 1942, p.490.
3. *Op.cit.*
4. Mining Industry Act, 1926, Coal Mines Act, 1930, Coal Act, 1938.
5. *The Economist,* 11 April 1942, p.490.
6. Coal Act, 1938.
7. *The Economist,* 5th February 1938, p.278.
8. *Op.cit.*
9. *Op. cit.*
10. For example, the port transport industry.
11. During the 1973-74 dispute there were frequent comments about the role of 'extremists' in the NUM.
12. See evidence to the Royal Commission on Nationalisation of Minerals discussed in Chapter 1.

TABLES

1. Output and Employment in the Coal Mining Industry, 1888-1913

Year	Numbers Employed (Thousands)	Output (Millions of Tons)
1888	439	170
1889	464	177
1890	507	182
1891	536	185
1892	549	182
1893	550	164
1894	570	188
1895	565	190
1896	557	195
1897	558	202
1898	567	202
1899	583	220
1900	624	225
1901	648	219
1902	663	227
1903	677	230
1904	682	232
1905	691	236
1906	710	251
1907	758	268
1908	796	262
1909	818	264
1910	848	264
1911	˙864	264
1912	879	260
1913	910	287

Source: J.W.F. Rowe, *Wages in the Coal Industry,* King, London, 1923, p.13.

2. Profits in the Coal Mining Industry, 1889-1913

Year	Amount : £ millions
Average 1889-1893	11.7
1894-1898	8.7
1899-1903	19.2
1904-1908	17.2
1907	24.4
1908	22.7
1909	14.9
1910	15.9
1911	15.3
1912	21.2
1913	28.0

Source: J.W.F. Rowe, *Wages in the Coal Industry*, King, London, 1923, p.170.

3.

Production of Coal and Profits in the Coal Mining Industry, 1924-1938

Year	Production of Coal Million Tons	Profits Per Ton Raised	
		s	d
1924	243	1	2
1925	221	0	3¼
1926	—	—	
1927	230	0	5¾ *
1928	219	0	11 *
1929	239	0	4½
1930	222	0	4½
1931	203	0	3½
1932	192	0	2
1933	191	0	2¾
1934	204	0	5
1935	206	0	6¼
1936	212	0	11½
1937	224	1	2¾
1938	210	1	4

* Losses made in these years.

Source: W.W. Haynes, *Nationalization in Practice: The British Coal Industry,* Bailey & Swinfen, London, 1953, p.28.

4.
Employment, Unemployment in the Coal Mining Industry, 1923-1938

Year	Persons Insured (Thousands)	Total Unemployed (Thousands)	Unemployment Rate Per Cent
1923	1,267	39	3
1924	1,259	72	6
1925	1,233	198	16
1926	1,226	–	–
1927	1,199	221	18
1928	1,116	252	23
1929	1,075	177	16
1930	1,069	219	20
1931	1,047	298	28
1932	1,045	355	34
1933	1,024	339	33
1934	982	281	28
1935	939	241	25
1936	896	199	22
1937	868	130	15
1938	858	133	16

Source: H. Wilson, *New Deal for Coal,* Cole, London, 1945, p.17.

5. Earnings in the Coal Mining Industry, 1923-1938

Year		Earnings Per Man Shift		Earnings: Quarterly Average		
		s	d	£	s	d
1923		10	1	33	11	6
1924		10	7¾	34	11	2
1925		10	6	32	18	11
1926	(Jan-April)	10	5	34	6	4
1927		10	0¾	30	13	1
1928		9	3½	28	9	2
1929		9	2½	29	11	7
1930		9	3½	28	9	7
1931		9	2¼	27	17	8
1932		9	2	27	7	1
1933		9	7½	27	11	6
1934		9	1¾	28	17	11
1935		9	3¼	29	12	0
1936		10	0¼	32	6	0
1937		10	8	36	0	10
1938		11	2¾	36	8	10

Source: R. Page Arnot, *The Miners: Years of Struggle,* Allen & Unwin, London, 1953, p.555.

6. Stoppages of Work in the Coal Mining Industry, 1926-1938

Year	Total Number of Stoppages			Number of Workers involved in Stoppages			Number of Man-Days Lost through Stoppages		
	All Industries (a)	Coal Mining Industry (b)	(b) as a percentage of (a) (c)	All Industries (d) 000s	Coal Mining Industry (e) 000s	(e) as a percentage of (d) (f)	All Industries (g) 000s	Coal Mining Industry (h) 000s	(h) as a percentage of (g) (i)
1926	323	1	0.3	2,751	1,050	38.2	162,233	145,209	89.5
1927	308	110	35.7	114	73	63.9	1,774	688	58.6
1928	302	97	32.1	124	82	66.1	1,388	452	32.6
1929	431	153	34.5	534	78	14.7	8,287	576	7.0
1930	422	150	35.5	309	149	48.1	4,399	663	15.1
1931	420	147	35.0	492	281	57.1	6,983	2,848	40.8
1932	389	111	28.5	382	52	13.7	6,488	287	4.4
1933	357	112	31.4	138	72	52.4	1,072	446	41.6
1934	471	143	30.4	134	73	54.7	959	365	38.0
1935	553	217	39.2	279	200	71.6	1,955	1,368	70.0
1936	818	270	33.0	322	182	56.4	1,829	852	46.6
1937	1,129	457	40.5	610	393	64.3	3,413	1,496	43.8
1938	875	363	41.5	275	174	63.1	1,334	697	52.2

Source: H. Wilson, *New Deal for Coal*, Cole, London, 1945, pp..36-7.

7. Estimated Loss of Coal Output Through Disputes, 1940-1944

Date	Tons
1940	500,600
1941	341,900
1942	833,200
1943	1,090,700
1944 1st Quarter	2,032,900
2nd Quarter	587,400
3rd Quarter	170,900
4th Quarter	210,500

Source: H. Wilson, *New Deal for Coal,* Cole, London, 1945, p.121.

8. Coal Consumption, Great Britain, 1947-1956

Million Tons

Inland Consumption	1947	1948	1949	1950	1951	1952	1953	1954	1955	1956
Electricity	27.1	28.8	30.0	33.0	35.4	35.7	36.7	39.7	42.9	45.6
Gas Works	22.7	24.6	25.3	26.2	27.4	27.8	27.2	27.4	27.9	27.8
Coke Ovens	19.8	22.3	22.6	22.6	23.5	25.2	25.9	26.7	27.0	29.3
Railways	14.3	14.3	14.4	14.2	14.2	13.9	13.4	13.1	12.2	12.1
Manufactured Fuel Plants	1.7	1.4	1.4	1.3	1.7	1.7	1.5	1.6	1.6	1.8
Collieries	11.1	11.3	10.8	10.7	10.6	10.3	9.9	9.5	8.6	7.9
Miners' Coal	5.0	5.0	5.0	5.0	5.3	5.3	5.2	5.3	5.1	5.2
Coastwise Bunkers	1.0	0.9	0.9	0.9	0.9	0.8	0.8	0.7	0.6	0.6
Iron & Steel Industry	8.4	8.4	8.2	8.1	7.8	7.7	7.2	6.7	6.5	6.1
Engineering Industry	2.7	3.0	3.0	3.2	3.3	3.4	3.3	3.5	3.6	3.4
Other Industry	24.7	26.5	27.2	28.8	29.9	27.8	29.1	30.7	30.6	29.9
Domestic	30.6	31.5	31.0	32.3	31.9	31.9	32.0	32.9	31.8	31.9
Miscellaneous	12.7	13.3	13.3	13.5	13.4	13.4	13.7	14.0	13.8	13.6
Total Inland Consumption	181.8	191.3	193.1	199.8	205.3	204.9	205.9	211.8	212.2	215.2
Shipments to N.Ireland	2.5	2.4	2.5	2.6	2.8	2.6	2.7	2.7	2.8	2.7
Overseas Shipments and Bunkers	5.5	16.3	19.4	17.1	11.7	15.1	16.9	16.3	14.1	10.0
Total Consumption and Shipments	189.8	210.0	215.0	219.5	219.8	222.6	225.5	230.8	229.1	227.9

Source: *Annual Abstract of Statistics*

9. Absence and Attendance in Coal Mining, 1946-1956

Year	Attendance: Manshifts Worked	Absence Percentage
1946	4.86	16.4
1947	4.72	12.5
1948	4.74	11.6
1949	4.70	12.4
1950	4.75	12.0
1951	4.84	12.2
1952	4.82	12.1
1953	4.73	12.4
1954	4.77	12.2
1955	4.74	12.5
1956	4.71	12.9

Source: L.J. Handy, "Absenteeism and Attendance in the British Coal Mining Industry : An Examination of Post War Trends", *British Journal of Industrial Relations,* Vol.VI, No.1, p.29.

10. Output Per Man Shift in N C B Mines, 1946-1956

Year	Tons Per Man Shift Face Workers	All Workers
1946	2.76	1.03
1947	2.86	1.07
1948	2.92	1.11
1949	3.02	1.16
1950	3.11	1.19
1951	3.18	1.21
1952	3.15	1.19
* 1953	3.22	1.22
1954	3.26	1.23
1955	3.26	1.23
1956	3.33	1.23

* New definitions of manpower and attendance were introduced in 1954.

The relevant figures for 1953 have been adjusted to the new basis.

Source: National Coal Board, *Report and Accounts for 1957,* H M S O, London, 1958, p.1.

11. Comparison of Average Male Earnings in Coal Mining and All Industries and Services, 1947-1956

Year	Coal Mining: Average Weekly Earnings (including allowances)		Coal Mining: Index of Weekly Earnings (1947 = 100)	All Industries and Services: Average Weekly Earnings		All Industries and Services: Index of Weekly Earnings (1947 = 100)
	s	d		s	d	
1947	147	1	100	120	9	100
1948	170	4	116	137	11	114
1949	181	11	124	142	8	118
1950	190	0	129	150	5	125
1951	212	10	145	166	0	137
1952	236	2	161	178	6	148
1953	244	6	166	189	2	157
1954	258	8	176	204	5	169
1955	274	6	187	217	5	180
1956	299	1	203	222	11	185

Source: *Annual Abstract of Statistics*

12. Stoppages of Work in Coal Mining, 1930-1956

Year	Number of Stoppages	Number of Workers Involved 000s	Number of Working days Lost 000s
1930	150	149	663
1931	147	281	2,848
1932	111	52	287
1933	112	72	446
1934	143	73	364
1935	217	200	1,368
1936	270	182	852
1937	457	393	1,494
1938	363	174	697
1939	404	206	565
1940	381	190	505
1941	463	153	334
1942	526	252	840
1943	835	295	890
1944	1,253	568	2,480
1945	1,295	242	640
1946	1,329	217	422
1947	1,049	308	912
1948	1,116	189	464
1949	872	248	754
1950	863	142	431
1951	1,058	135	350
1952	1,221	274	660
1953	1,307	168	393
1954	1,464	204	468
1955	1,783	354	1,112
1956	2,067	241	502

Source: *Ministry of Labour Gazette*

13. Tonnage Lost Per Man in Disputes in the Coal Mining Industry, 1947-1956, Coalfield Comparison

Year	Tonnage Lost Per Man Through Disputes:							
	Scotland	Northumberland & Cumberland	Durham	Yorkshire	N. Western	E. Midlands	W. Midlands	S. Western
1947	4.62	0.73		6.28	0.23	0.32	0.10	0.95
1948	2.41	0.12		3.33	0.70	0.57	0.30	0.68
1949	0.03	*	0.30	2.80	0.70	5.38	0.34	0.74
1950	0.01	*	0.10	1.33	0.99	12.60	0.29	0.59
1951	3.08	0.09	0.13	2.04	0.76	0.51	0.13	1.40
1952	4.39	0.10	0.46	4.54	0.64	0.32	0.19	2.03
1953	4.42	0.08	0.11	2.35	0.35	0.29	0.44	1.29
1954	5.11	0.10	0.62	3.42	0.40	0.27	0.64	1.43
1955	5.43	0.15	0.11	12.22	1.32	0.47	0.36	2.94
1956	4.73	0.21	0.16	5.24	1.31	2.01	0.90	2.89

South Eastern excluded for reasons of size.

* Less than 0.01

Source: B.J. McCormick, "Strikes in the Yorkshire Coalfield, 1947-1963", *Economic Studies*, Vol.4, p.179.

14. Coal Production and Consumption, UK, 1956-1970

Million Tons

Year	Coal Production, UK			Coal Consumption		
	Deep Mined	Open Cast	Total	Inland	Export and Bunkers	Total
1956	209.9	12.1	222.0	217.5	9.7	227.2
1957	210.0	13.6	223.6	212.9	7.9	220.8
1958	201.5	14.3	215.8	202.4	4.9	207.3
1959	195.3	10.8	206.1	189.4	4.3	193.7
1960	186.0	7.6	193.6	196.7	5.5	202.2
1961	181.9	8.6	190.5	191.8	5.7	197.5
1962	189.3	8.1	197.4	191.2	4.8	196.0
1963	189.7	6.1	195.8	194.0	7.5	201.5
1964	186.8	6.8	193.6	187.2	6.0	193.2
1965	180.2	7.3	187.5	184.6	3.7	188.3
1966	167.6	7.0	174.6	174.7	2.8	177.5
1967	165.0	7.1	172.1	163.8	1.9	165.7
1968	157.2	6.9	164.1	164.5	2.7	167.2
1969	144.2	6.3	150.5	161.1	3.5	164.6
1970	134.5	7.8	142.3	154.4	3.2	157.6

Source: *Annual Abstract of Statistics*

15. Coal Consumption, UK, by Industry, 1957-1970

Million Tons

Industry etc.	1956	1957	1958	1959	1960	1961	1962	1963	1964	1965	1966	1967	1968	1969	1970
Electricity Supply	45.6	46.4	46.1	46.0	51.1	54.7	60.4	66.8	67.4	69.3	68.6	67.2	73.2	75.9	76.0
Gas Supply	27.8	26.4	24.8	22.5	22.3	22.2	22.1	22.1	20.2	18.0	16.9	14.6	10.7	6.9	4.2
Coke Ovens	29.3	30.7	27.8	25.7	28.5	26.8	23.5	23.5	25.5	25.7	24.8	23.6	24.9	25.3	24.9
Manufactured Fuel Plants	1.8	2.1	2.2	1.7	1.4	1.5	1.6	1.7	1.4	0.9	1.7	2.0	2.2	2.4	2.6
Railways	12.1	11.4	10.3	9.5	8.9	7.7	6.1	4.9	3.8	2.8	1.7	0.8	0.2	0.2	0.1
Collieries	7.9	7.2	6.5	5.6	5.0	4.5	4.2	3.9	3.7	3.4	3.1	2.9	2.4	2.0	1.9
Iron and Steel	5.8	5.3	4.2	3.7	3.8	3.2	2.6	2.4	2.1	1.8	1.3	0.9	0.8	0.9	0.8
Engineering and Other Metal Trades	3.7	3.4	3.3	3.0	2.9	2.8	2.8	2.7	2.5	2.5					
* Other Industry	30.0	28.8	25.0	24.6	23.2	22.0	20.8	20.1	19.8	23.9	22.0	21.8	20.5	18.5	
House Coal	30.6	28.8	26.9	28.0	26.1	26.4	25.7	22.0	21.5	20.9	18.8	18.1	16.8	15.2	
Anthracite & Dry Steam Coal (domestic)	1.6	1.5	1.5	1.5	1.6	1.6	1.6	1.7	1.6	1.8	1.8	1.7	1.9	1.9	2.0
Miners' Coal	5.3	5.3	5.0	4.9	4.7	4.7	4.6	4.3	4.2	3.8	3.6	3.2	2.9	2.7	
Miscellaneous	13.0	12.6	11.9	10.6	10.9	10.0	10.4	10.4	10.1	10.4	6.9	6.5	6.0	6.4	6.6
Shipments to N.Ireland	2.7	2.8	2.7	2.5	2.7	2.6	2.6	2.6	2.3	2.5					
Shipments to Channel Is.	0.3	0.2	0.2	0.2	0.2	0.2	0.2	0.2	0.2	0.2					
†TOTAL	217.5	212.9	202.4	189.4	196.7	191.8	191.2	194.0	187.2	184.6	174.7	163.8	164.5	161.1	154.4

* Includes Engineering and Other Metal Trades after 1965

† Excludes Shipments to N. Ireland and Channel Islands after 1965.

Source: *Annual Abstract of Statistics.*

16. UK Energy Consumption, 1960-1970

Million Therms

Type of Fuel	1960	1961	1962	1963	1964	1965	1966	1967	1968	1969	1970
Coal (Direct Use)	23,433	21,762	20,983	20,060	18,048	17,409	16,122	14,402	13,613	12,883	11,839
%	46	43	41	38	35	32	30	27	25	23	20
Coke & Breeze	7,118	6,749	6,525	6,621	6,713	6,653	6,032	5,636	5,740	5,553	5,141
%	14	14	13	13	13	12	11	11	10	10	9
Other Solid Fuel	588	601	618	655	619	572	623	667	767	856	867
%	1	1	1	1	1	1	1	1	1	2	1
Coke Oven Gas	551	517	440	447	523	530	483	461	483	452	462
%	1	1	1	1	1	1	1	1	1	1	1
Town Gas	2,636	2,618	2,715	2,922	3,026	3,338	3,685	3,983	4,352	4,536	4,289
%	5	5	5	6	6	6	7	7	8	8	7
Natural Gas	—	—	—	—	—	—	—	1	82	426	1.454
%	—	—	—	—	—	—	—	0	0	1	3
Electricity	3,372	3,636	4,046	4,432	4,653	5,022	5,222	5,391	5,820	6,245	6,554
%	7	7	8	8	9	9	10	10	11	11	11
Petroleum	12,385	13,614	15,133	16,799	18,159	19,820	21,034	22,492	24,063	25,628	27,198
%	25	27	30	32	35	37	39	42	44	45	47
Other Fuels	448	400	439	437	369	343	364	320	249	189	159
%	1	1	1	1	1	1	1	1	0	0	0
Total	50,531	49,861	50,899	52,373	52,110	53,687	53,565	53,353	55,169	56,768	57,963
%	100	100	100	100	100	100	100	100	100	100	100

Source: *Annual Abstract of Statistics*

17. Numbers Employed in Coal Mining, 1957-1971

Year	Number of Wage Earners Employed 000s
1957	703.8
1958	692.7
1959	658.2
1960	602.1
1961	570.5
1962	550.9
1963	523.8
1964	497.8
1965	465.6
1966/7	419.4
1967/8	391.9
1968/9	336.3
1969/70	305.1
1970/71	287.2

Source: *Annual Abstract of Statistics*

18. Pits Operated by N C B, 1956-1970

Year	Number of Pits
1956	840
1957	822
1958	793
1959	737
1960	698
1961	669
1962	616
1963/4	576
1964/5	534
1965/6	483
1966/7	438
1967/8	476
1968/9	317
1969/70	299

Source: E.F. Schumacher, "Some Aspects of Coal Board Policy 1947-1967", *Economic Studies,* Vol.V, No.1, 1969, p.18; National Board for Prices and Incomes, *Coal Prices* (2nd Report), Report No. 153, Cmnd. 4455, H M S O, London, 1970, p.5.

19. Mechanisation in the British Coal Mining Industry, 1961-1969

Year	Number of Longwall Faces Operating During Month of September (Excluding Fully Reserved Training Faces)				
	Mechanised		Non-Mechanised		Total
	No.	Per Cent	No.	Per Cent	No.
1961	1,147	38.0	1,875	62.0	3,022
1962	1,255	45.9	1,478	54.1	2,733
1963	1,345	54.4	1,128	45.6	2,473
1964	1,415	61.4	889	38.6	2,304
1965	1,389	67.1	681	32.9	2,070
1966	1,357	76.4	419	23.6	1,776
*1967	1,323	82.5	280	17.5	1,603
	1,340	83.5	264	16.5	1,604
1968	1,066	84.4	197	15.6	1,263
1969	924	86.8	141	13.2	1,065

*Since 1967 the number of faces operating has been collected for one sample week in the month. Faces which were previously counted as mechanised if they were operating as such for at least half of the month, are now counted as mechanised if operating as such for any part of the sample week.

Source: National Board for Prices and Incomes, *Coal Prices* (2nd Report), Report No. 53, Cmnd. 4455, H M S O, London, 1970, p.4.

20. Productivity in Coal Mining, 1957-1971

Year	Output Per Manshift (Tons)
1957	1.25
1958	1.28
1959	1.35
1960	1.40
1961	1.45
1962	1.56
1963	1.65
1964	1.72
1965	1.80
1966/7	1.83
1967/8	1.95
1968/9	2.12
1969/70	2.17
1970/71	2.21

Source: *Annual Abstract of Statistics*

21. Comparison of Miners' Earnings with those in Other Industries, 1956-1970

Year	Average Weekly Earnings in Coal Mining	Average Weekly Earnings in Manufacturing Industries	Coal-mining as a percentage of Manufacturing
	£	£	
1956	15.35	12.28	125
1957	16.32	13.06	125
1958	15.41	13.27	116
1959	15.70	14.21	111
1960	16.28	15.16	107
1961	17.16	15.89	108
1962	17.93	16.34	110
1963	18.75	17.29	108
1964	19.73	18.67	106
1965	21.21	20.16	105
1966	22.16	20.78	107
1967	22.92	21.89	105
1968	24.12	23.62	102
1969	25.10	25.54	98
1970	28.01	28.91	97

The figures for coal-mining, provided by the N C B, include allowances for holiday and rest pay and for sickness pay, but exclude the value of allowances in Kind.

Source: Pay Board, *Special Report: Relative Pay of Mineworkers,* Cmnd. 5567, H M S O, London, 1974. p.31.

22. Stoppages of Work in the Coal Mining Industry, 1957-1970

Year	Number of Stoppages	Numbers of Workers involved in Stoppages 000s	Number of Working Days Lost Through Stoppages 000s
1957	2,224	265	514
1958	1,963	249	450
1959	1,307	191	363
1960	1,666	237	494
1961	1,458	249	737
1962	1,203	154	308
1963	987	152	326
1964	1,059	168	302
1965	731	117	412
1966	553	50	118
1967	394	41	105
1968	215	29	53
1969	193	145	1,041
1970	165	118	1,092

Source: *Ministry of Labour Gazette, Department of Productivity Gazette, Department of Employment and Productivity Gazette.*

23. Stoppages of Work in the Coal Mining Industry Weighted by Number of Miners Employed, 1957-1970

Year	Number of Stoppages per 1,000 workers	Number of Workers Involved in Stoppages per 1,000 workers	Number of Working Days Lost Through Stoppages per 1,000 workers
1957	3.16	376.95	730.11
1958	2.84	359.82	649.35
1959	1.99	290.27	551.67
1960	2.77	393.69	820.60
1961	2.55	463.84	1,290.72
1962	2.18	279.49	558.98
1963	1.88	290.08	622.14
1964	2.13	338.03	606.43
1965	1.57	251.07	884.12
1966	1.32	119.33	281.62
1967	1.01	104.59	267.86
1968	0.64	86.31	157.74
1969	0.63	475.41	3,413.11
1970	0.57	411.15	3,804.88

Source: *Ministry of Labour Gazette, Department of Productivity Gazette, Department of Employment and Productivity Gazette.*

24. Coal Production and Consumption, UK, 1969-1972

| | Million Tons | | | |
	1969	1970	1971	1972
Production:				
Deep Mined	144.2	134.5	134.3	107.3
Open Cast	6.3	7.8	10.5	10.3
Total	164.1	142.3	144.8	117.6
Imports	0	0.1	4.2	4.9
Consumption:				
Home	161.1	154.4	138.7	120.9
Overseas and Bunkers	3.5	3.2	2.6	1.8
Total	164.6	157.6	141.3	122.7

Source: *Annual Abstract of Statistics*

25. Movement in Relative Weekly Earnings of Full-Time Men Employed in Coal Mining, 1960-1971

Date	Industrial Group: Manufacturing	Coal Mining	Coal Mining as a ratio of Manufacturing
October 1960			
Average Weekly Earnings	£15.16	£16.28	107.4
Rank		3rd	
October 1965			
* Average Weekly Earnings	£20.16	£21.21	105.2
Rank		3rd	
October 1968			
Average Weekly Earnings	£22.82	£24.12	105.7
Rank		5th	
October 1970			
Average Weekly Earnings	£28.91	£28.01	96.9
Rank		12th	
October 1971			
Average Weekly Earnings	£31.36	£31.65	100.9
Rank		9th	

(Out of 21 industry groups.)

* J. Hughes and R. Moore claim that these figures are incorrect and should read £23.62 and 102.1. See J. Hughes and R. Moore (eds.), *A Special Case?*, Penguin, Harmondsworth, 1972, p.128.

Source: Department of Employment, *Report of a Court of Inquiry into a Dispute Between the National Coal Board and the National Union of Mineworkers* (Chairman Wilberforce), Cmnd. 4903, H M S O, London, 1972.

26. Posted Petroleum Prices, 1972-1974

	20 January 1972	1 January 1973	Dollars per barrel 16 October 1973	1 January 2974
Persian Gulf				
Arabian Light (34°)	2.479	2.591	5.119	11.651
Abu Dhabi Murban (39°)	2.540	2.654	6.045	12.630
Mediterranean and African				
Arabian Light (34°)	3.370	3.451	7.149	13.647
Libyan (40°)	3.673	3.777	8.925*	15.768
Nigerian (34°)	3.446†	3.561	8.310	14.691
Venezuela				
Oficina (35°)	3.261	3.477	7.802ˣ	14.247ᶿ

* 19 October   † 15 February   x 1 November   ᶿ Excluding sulphur premium

Source: *The Petroleum Economist*, Vol.XL1, No.2, February 1974, p.43.

27. Result of 1974 Strike Ballot

	Total Votes	Total Voting Yes	Percentage Voting Yes
Yorkshire	54,570	49,278	90.30
Nottinghamshire	28,284	21,801	77.08
South Wales	29,901	25,058	93.15
Durham	17,341	14,861	85.70
C.O.S.A.	15,368	6,066	39.47
Scotland	16,587	14,497	87.40
Midlands	12,309	9,016	73.25
Derbyshire	10,679	9,242	86.54
North West	8,637	7,084	82.02
Northumberland	8,420	7,575	84.03
Durham Mechanics	5,937	4,590	77.31
Group No 2 (Scotland)	4,834	3,929	81.28
Cokemen	4,583	3,076	67.12
Power Group	3,981	2,239	56.24
South Derbyshire	2,604	1,827	70.16
Leicestershire	2,519	1,553	61.65
Kent	2,360	2,117	89.70
Northumberland (Mechanics)	2,191	1,816	82.88
North Wales	1,200	952	79.33
Cumberland	880	775	88.07
Power Group No 2	1,164	681	58.51
Durham Enginemen	896	543	60.60
Yorkshire Enginemen	370	316	85.41
Total	232,615	188,393	80.99

Source: *Guardian*, 5 February 1974, p.5.

BIBLIOGRAPHY

Books, Pamphlets etc.

Action Society Trust, Nationalised Industry Series:
1. *Accountability to Parliament,* 1950;
2. *The Powers of the Minister,* 1951;
6. *The Extent of Centralisation, Part I,* 1951;
7. *The Extent of Centralisation, Part II,* 1951;
11. *The Workers' Point of View,* 1952;
Acton Society, London.

Arnot, R. Page, *Further Facts From the Coal Commission,* Labour Research Department, London, 1920.

Arnot, R. Page, *The Miners; A History of the Miners' Federation of Great Britain, 1889-1910,* Allen & Unwin, London, 1949.

Arnot, R. Page, *The Miners in Crisis and War,* Allen & Unwin, London, 1961.

Arnot, R. Page, *The Miners: Years of Struggle,* Allen & Unwin, London, 1953.

Baldwin, G.B., *Beyond Nationalisation – The Labour Problems of British Coal,* Harvard University Press, 1955.

Barry, E.E., *Nationalisation in British Politics: The Historical Background,* Jonathan Cap, London, 1965.

Benney, M., *Charity Main: A Coalfield Chronicle,* Allen & Unwin, London, 1946.

Brown, W., *Piecework Abandoned,* Heinemann, London, 1962.

Burn, D. (ed.), *The Structure of British Industry,* Cambridge University Press, London, 1958.

Chester, D.N., *The Nationalised Industries: An Analysis of the Statutory Provisions,* Allen & Unwin, London, 1951.

Clegg, H., *Labour in Nationalised Industry,* Fabian Society, London, 1950.

Coates, D., and Topham, T., *The New Unionism,* Peter Owen, London, 1972.

Cole, G.D.H., *Labour in the Coal Mining Industry, 1914-1921,* Clarendon Press, Oxford, 1923.

Cole, G.D.H., *National Coal Board,* Fabian Society, London, 1949 (Revised edition), Fabian Research Series, 129.

Cole, M., *Miners and the Board,* Fabian Society, London, 1949, Fabian Society, London, 1949, Rabian Research Series, 134.

Cook, A.J., *The Coal Shortage: Why the Miners Will Win,* Labour Research Department, London, 1926.

Court, W.H., *Coal,* Longmans, Green & Co., London, 1951.

Dennis, N, Henriques, F, and Slaughter, C, *Coal is Our Life,* Eyre and Spottiswoode, London, 1956.

Department of Employment and Productivity, *Ryhope: A Pit Closes,* H M S O, London, 1970.

Dickie, J.P., *The Coal Problem, A Survey: 1910-1936,* Methuen, London, 1936.

Ditz, G.W., *British Coal Nationalised,* Edward Hazen Foundation, New Haven, 1951.

Dron, G.W., *The Economics of Coal Mining,* Arnold, London, 1928.

Dunning, J.H., and Thomas, C.J., *British Industry,* Hutchinson, London, 1966.

Evans, E.W., *The Miners of South Wales,* University of Wales Press, Cardiff, 1961.

Fabian Group, *Plan for Coal Distribution,* Fabian Society, London, 1956, Fabian Research Series, 182.

Foot, R., *Plan for Coal,* Mining Association of Great Britain, London, 1945.

Gibson, W., *Coal in Great Britain,* Arnold, London, 1920.

Gregory, R., *The Miners and British Politics, 1906-1914,* Oxford University Press, London, 1968.

Hancock, W.K., and Gowing, M.M., *British War Economy,* H M S O, London, 1969.

Haynes, W.W., *Nationalization in Practice: The British Coal Industry,* Bailey & Swinfen, London, 1953.

Heinemann, M., *Britain's Coal: A Study of the Mining Crisis,* Victor Gollancz, London, 1944.

Heinemann, M., *Coal Must Come First,* Frederick Muller, London, 1948.

Hodges, F., *Nationalisation of the Mines,* Leonard Parsons, London, 1920.

Hogg, Q., Lancaster Col., Thorneycroft P., (eds.), *Forward by the Right: A National Policy for Coal,* Tory Reform Committee, London, 1944.

House, J.W. and Knight, E.M., *Pit Closure and the Community,* University of Newcastle, Department of Geography, 1967.

Hughes, J. and Moore, R., (eds), *A Special Case?* Penguin, Harmondsworth, 1972.

Jevons, W.C., *The Coal Question,* Macmillan, London, 1906 (3rd edition).

Kelf-Cohen, R., *Nationalisation in Britain,* Macmillan, London, 1959.

Kelly, D.M., and Forsyth, D.J.C. (eds.), *Studies in the British Coal Industry,* Pergamon, Oxford, 1969.

Knight, E.M., *Men Leaving Mining,* University of Newcastle, Department of Geography, 1968.

Lancaster, Col. C.G., *About Coal,* Conservative Political Centre, London, 1948.

Lancaster, Col. C.G., Reid, Sir C., Young, Sir C., *Structure and Control of the Coal Industry,* Conservative Political Centre, London, 1951.

Lee, W.A., *Thirty Years in Coal,* Mining Association of Great Britain, London, 1954.

Lubin, I. and Everett, H., *British Coal Dilemma,* Macmillan, New York, 1927.

Moffat, A., *My Life with the Miners,* Lawrence & Wishart, London, 1965.

Moore, R., *Pitmen, Preachers and Politics,* Cambridge University Press, London, 1974.

National Coal Board, *Investing in Coal,* N C B, London, 1956.

National Coal Board, *Plan for Coal,* N C B, London, 1950.

National Coal Board, *Revised Plan for Coal,* N C B, London, 1959.

Nef, J.U., *The Rise of the British Coal Industry,* George Routledge & Sons, London, 1932.

Neuman, A.M., *Economic Organisation of the British Coal Industry,* George Routledge & Sons, London, 1934.

Platt, J., *British Coal: A Review of the Industry, its Organisation and Management,* Lyon, Grant and Green, London, 1968.

Political and Economic Planning, *Report on British Coal Industry,* P E P, London, 1936.

Political and Economic Planning, *The British Fuel and Power Industries,* P E P, London, 1947.

Pollard, S., *The Development of the British Economy,* Edward Arnold, London, 1962.

Raynes, J.R., *Coal and its Conflicts: A Brief Record of the Disputes Between Capital and Labour in the Coal Mining Industry of Great Britain,* Ernest Benn, London, 1928.

Redmayne, R.A.S., *Men, Mines and Memories,* Eyre and Spottiswoode, London, 1942.

Redmayne, R.A.S., *The British Coal-Mining Industry During the War,* Clarendon Press, Oxford, 1923.

Redmayne, R.A.S., *The Problem of the Coal Mines,* Eyre and Spottiswoode, London, 1945.

Reid, G.L. and Allen, K., *Nationalised Industries,* Penguin, Harmondsworth, 1970.

Reid, G.L., Allen, K., and Harris, D.J., *The Nationalized Fuel Industries,* Heinemann, London, 1973.

Robinson, C., *A Policy for Fuel?* Institute of Economic Affairs, London, 1969.

Robinson, C., *Competition for Fuel*, Institute of Economic Affairs, London, 1971.

Robson, W.A., *Nationalised Industry and Public Ownership*, Allen & Unwin, London, 1962.

Rowe, J.W.F., *Wages in the Coal Industry*, King, London, 1923.

Scott, W.H., *et al., Coal and Conflict*, Liverpool University Press, Liverpool, 1963.

Simpson, E.S., *Coal and Power Industries in Post War Britain*, Longmans, London, 1966.

Smart, R.C., *The Economics of the Coal Industry*, King, London, 1930.

Thomas, I., *Coal in the New Era*, Putnam, London, 1934.

Townsend-Rose, H., *The British Coal Industry*, Allen & Unwin, London, 1951.

Unofficial Reform Committee, *The Miners Next Step: Being A Scheme for the Reorganisation of the Federation*, Pluto Press, 1973. (reprint).

White, E., *Workers' Control?* Fabian Society, London, 1951 (revised edition), Fabian Tracts, 271.

Williams, J.E., *The Derbyshire Miners*, Allen & Unwin, London, 1962.

Wilson, H., *New Deal for Coal*, Cole, London, 1945.

Youngson, A.J., *The British Economy, 1920-1957*, Allen & Unwin, London, 1962.

Zweig, F., *Men in the Pits*, Victor Gollancz, London, 1949.

Reports etc.

Board of Trade, *Reports of Departmental Committee on Co-operative Selling in the Coal Mining Industry* (Chairman Lewis), Cmd. 2770, H M S O, London, 1926.

Board of Trade, *Report on the Working of Part I of the Mining Industry Act, 1926,* Cmd. 3214, H M S O, London, 1928.

Board of Trade,*Memorandum on the Production, Distribution and Rationing of Coal,* Cmd. 6364, H M S O, London, 1942.

Board of Investigation,
1. *The Immediate Wages Issue,* H M S O, London, 1942;
2. *Supplemental Report: Output Bonus,* H M S O, London, 1942;
3. *Third Report: Machinery for Determining Wages and Conditions of Employment,* H M S O, London, 1943;
4. *Fourth and Final Report,* H M S O, London, 1943.
(ChairmanGreene).

Coal Industry Commission, *Interim Report,* Cmd. 359, H M S O, London, 1919; *Final Report,* Cmd. 361, H M S O, London, 1919. (Chairman Sankey).

Coal Industry Commission, *Report* (Chairman Samuel) Cmd. 2600, H M S O, London, 1926.

Coal Mines Reorganisation Commission, Reports on the Working of the 1930 Act, 1. Cmd. 4468, H M S O, London, 1933; 2. Cmd. 5069, H M S O, London, 1936.

Department of Economic Affairs, *National Plan,* Cmnd. 2704, H M S O, London, 1965.

Department of Employment, *Report of a Court of Inquiry into a Dispute Between the National Coal Board and the National Union of Mineworkers* (Chairman Wilberforce), Cmnd. 4903, H M S O, London, 1972.

Economic Survey for 1947, Cmd. 7046, H M S O, London, 1947.

Economic Survey for 1948, Cmd. 7344, H M S O, London, 1948.

Financial and Economic Obligations of Nationalised Industries, Cmnd. 1337, H M S O, London, 1961.

Financial Position of the Coal Mining Industry: Coal Charges Account, Cmd. 6617, H M S O, London, 1945.).

Ministry of Fuel and Power, *Recruitment of Juveniles in the Coal Mining Industry* (Chairman Foster), H M S O, London, 1942.

Ministry of Fuel and Power, *Report of Technical Advisory Committee on Coal Mining,* (Chairman Reid) Cmd. 6610, H M S O, London, 1945.

Ministry of Fuel and Power, *Coal Industry Nationalisation,* Cmd. 6716, H M S O, London, 1945.

Ministry of Fuel and Power, *Report of the Committee on National Policy for the Use of Fuel and Power Resources,* (Chairman Ridley), Cmd. 8647, H M S O, London, 1952.

Ministry of Labour, *Report of a Committee of Inquiry Concerning the Wages Position in the Coal Mining Industry,* (Chairman Buckmaster), H M S O, Cmd. 2129, London, 1924.

Ministry of Labour, *Report of a Committee of Inquiry Concerning the Coal Mining Industry Dispute,* (Chairman MacMillan), Cmd. 2478, H M S O, London, 1925.

Ministry of Labour, *Report of the Committee of Investigation into Differences Between the Yorkshire Winding Enginemen's Association and the Members of the N U M Employed by the N C B* (Chairman Wilson), H M S O, London, 1964.

Ministry of Labour, *Report of an Inquiry into the Differences in the South Wales Coalfield* (Chairman Picton), H M S O, London, 1965.

Ministry of Power, *Report of the Court of Inquiry into "Coal Distribution Costs in Great Britain"* (Chairman Robson), Cmnd. 446 , H M S O, London, 1958.

Ministry of Power, *The Finances of the Coal Industry,* Cmnd. 2805, H M S O, London, 1965.

Ministry of Power, *Fuel Policy,* Cmnd. 2798, H M S O, London, 1965.

Ministry of Power, *Fuel Policy,* Cmnd. 3438, H M S O, London, 1967.

National Board for Prices and Incomes, *Coal Prices,* Cmnd. 2919,
H M S O, London, 1966, Report No. 12.

National Board for Prices and Incomes, *Coal Distribution Costs,* Cmnd.
3094, H M S O, London, Report No. 21.

National Board for Prices and Incomes, *Coal Prices,* Cmnd.4149,
H M S O, London, 1969, Report No. 124.

National Board for Prices and Incomes, *Coal Prices,* (2nd Report), Cmnd.
4455, H M S O, London 1970, Report No. 153. Supplement No. 1,
Cmnd. 4455-1, H M S O, London, 1971, Supplement No.2, Cmnd.
4455-2, H M S O, London, 1971.

National Tribunal for the Coal Mining Industry, *Fifth Report* (Chairman
Porter), H M S O, London, 1944

Pay Board, *Problems of Pay Relativities,* Cmnd. 5535, H M S O, London,
1974.

Pay Board, *Special Report: Relative Pay of Mineworkers,* Cmnd. 5567,
H M S O, London, 1974.

Select Committee on Nationalised Industries, *Reports and Accounts,*
H M S O, London, 1958.

Select Committee on Nationalised Industries, *Exploitation of North Sea
Gas,* H M S O, London, 1968.

Select Committee on Nationalised Industries, *Nationalised Coal Board,*
H M S O, London, 1969.

Articles

Alexander, K.J.W., "Wages in Coal Mining Since Nationalisation",
Oxford Economic Papers, Vol.8, No. 2,1956, pp. 164-81.

Barratt-Brown, M., "Coal as a Nationalised Industry", *Economic Studies,*
Vol.4, No.1/2, October 1969, pp.95-134.

Beacham, A., "The Coal Industry in Great Britain", *The Economic
Journal,* Vol.LX, 1950, pp. 9-18.

Burns, D., "The National Coal Board", *Lloyd's Bank Review,* Vol.43,
1951, pp. 33-48.

Clegg, H.A., "The Fleck Report", *Public Administration,* Vol.XXXIII,
Autumn 1955, pp. 274-5.

Cole, G.D.H., "The National Coal Board", *Political Quarterly,* October-
December 1946, Vol. 17, No.4, pp. 310-19.

Edwards, E., "A National Union for All Mine Workers in Great Britain",
Colliery Guardian, Vol. CLIX, 17 November 1944, pp. 568-9.

Handy, L.J., "Absenteeism and Attendance in the British Coal Mining
Industry: An Examination of Post War Trends", *British Journal of
Industrial Relations,* Vol. VI, No.1, pp. 27-50.

Hanson, A.H., "Labour and the Public Corporation", *Public Administration*, Vol.32, 1954, pp. 203-9.

Hanson, A.H., "Parliamentary Question on the Nationalised Industries", *Public Administration*, Vol.XXIX, Spring 1951, pp.51-62.

Hepworth, R., *et al.*, "The Effects of Technological Change in the Yorkshire Coalfield, 1960-1965", *Economic Studies*, Vol.4, No. 1/2, October 1969, pp. 221-38.

Kelly, D.M., "Contraction of the Coal Industry: Some Aspects of the Effects on Manpower", *Economic Studies*, Vol.4, No. 1/2, October 1969, pp.199-220.

Kerr, C., and Siegel, A., "The Inter-Industry Propensity to Strike – An International Comparison", in Flanders, A. (eds.), *Collective Bargaining*, Penguin, Harmondsworth, 1969.

Liddell, F.D.K., "Attendance in the Coal Mining Industry", *British Journal of Sociology*, Vol.5, 1954, pp. 78-86.

McCormick, B.J., "Management Unionism in the Coal Industry", *British Journal of Sociology*, Vol.XI, 1960, pp. 356-69.

McCormick, B.J., "Strikes in the Yorkshire Coalfield, 1947-1963", *Economic Studies*, Vol.4, 1969, pp.171-97.

Moos, S., "The Statistics of Absenteeism in Coal Mining", *Manchester School for Economic and Social Studies*, Vol.19, January, 1951, pp. 89-108.

Patterson, T.T., and Willett, F.J., "Unofficial Strike", *Sociological Review*, Vol. XLIII, 1951, pp.57-94.

Prest, W., "The British Coal Mines Act of 1930, Another Interpretation", *Quarterly Journal of Economics*, Vol.L, February 1936, pp.136-49.

Reid, C., "The Problem of Coal", The *Times*, 22-24 November 1948.

Rimlinger, G.V., "National Differences in the Strike Propensity of Coal Miners: Experience in Four Countries", *International and Labour Relations Review*, Vol.12, No.3, April 1959, pp.389-406.

Schumacher, E.F., "Some Aspects of Coal Board Policy, 1947-1967", *Economic Studies*, Vol.V, No.1, 1969, pp.3-30.

Slaughter, C., "The Strike of the Yorkshire Mineworkers in May, 1955", *Sociological Review*, Vol.6, 1958, pp.241-59.

Snaith, W., "The Adjustment of the Labour Force on the Durham Coalfield – A Study of Redundancy", *Economic Studies*, Vol.4, No.1/2, October 1969, pp.239-52.

Tawney, R.H., "The British Coal Industry and the Question of Nationalisation", *Quarterly Journal of Economics*, Vol.35, November 1920, pp.61-107.

Trist, E.L., and Bamforth, K.W., "Some Social and Psychological

Consequences of the Longwall Method of Coal-Getting", *Human Relations,* Vol.4, 1951, pp.3-38.

Vaughan, G., "What the Miners Think", *Coal,* September 1949, p.10.

Wellisz, S., "Strikes in Coal Mining", *British Journal of Sociology,* Vol.IV, 1956, pp.346-66.

Theses

Broadley, O., *The Colliery Consultative Committee,* M A thesis, University of Liverpool, 1959.

Griffin, A.R., *The Development of Industrial Relations in a Nottinghamshire Coalfield,* Ph.D. thesis, Nottingham, 1963.

Hopkins, T.N., *The Operation of the National Reference Tribunal in the Coal Industry Since 1943,* MA thesis, Aberystwyth, 1961.

Jackson, M.P., *A Critical Analysis of the Ministry of Labour's Method of Classifying the Causes of Stoppages, with Special Reference to Major Stoppages in the Port Transport and Coal Mining Industries Between 1963 and 1966,* MA thesis, Hull, 1971.

Kelly, D.W., *The Study of the Administration of the Coal Industry in Great Britain Since 1946 with Special Reference to the Problem of Decentralisation,* MA thesis, Wales, 1952.

Smith, C.S., *Planned Transfer of Labour with Special Reference to the Coal Industry,* Ph.D. thesis, Bedford College, 1960.

Statutes etc.

Armed Forces (Conditions of Service) Act, 1939.
Coal Act, 1938.
Coal Industry Act, 1965.
Coal Industry Act, 1967.
Coal Industry Act, 1971.
Coal Industry Act, 1973.
Coal Mines Act, 1930.
Coal Mines (Emergency) Act, 1920.
Coal Mines Regulation Act, 1908.
Coal Nationalisation Act, 1946.
Control of Employment Act, 1939.
Defence of the Realm Act, 1914.
Essential Work (Coalmining Industry) Order, 1941.
Fuel and Lighting Order, 1939.
Minimum Wage Act, 1912.
Mining Industry Act, 1920.
Mining Industry Act, 1926.
Munitions of War Act, 1915.
National Service (Armed Forces) Act, 1939.
Price of Coal (Limitation) Act, 1915.
Registration for Employment Order, 1941.

Statistical Sources

Annual Abstract of Statistics.
Coal.
Coal Figures.
Department of Employment and Productivity Gazette.
Employment and Productivity Gazette.
Ministry of Labour Gazette.
Ministry of Power Statistical Digest.
National Coal Board, *Annual Report and Accounts.*

Index

214